Ecological Studies, Vol. 115

Analysis and Synthesis

Edited by

O.L. Lange, Würzburg, FRG
H.A. Mooney, Stanford, USA

Ecological Studies

Volumes published since 1989 are listed at the end of this book.

Springer
Berlin
Heidelberg
New York
Barcelona
Budapest
Hong Kong
London
Milan
Paris
Tokyo

Peter M. Vitousek   Lloyd L. Loope
Henning Adsersen (Eds.)

Islands

Biological Diversity and Ecosystem Function

With 17 Figures

Springer

Prof. Dr. Peter M. Vitousek
Department of Biological Sciences
Stanford University
Stanford, CA 94305, USA

Dr. Lloyd L. Loope
National Biological Survey
Haleakala National Park
Makawao, Maui, HI 96768, USA

Prof. Dr. Henning Adsersen
Department of Plant Ecology
Botanical Institute
University of Copenhagen
Øster Farimagsgade 2D
DK1353 Copenhagen K, Denmark

Front cover: Illustration by Kirsten Madsen
Famous examples of island biodiversity: the dodo, a giant flightless dove, evolved and became extinct on Mauritius; the sea coconut, producer of the world's biggest seed, endemic to the Seychelles, possibly a relic from a sunken subcontinent; and the Darwin finches of the Galapagos Islands, inspiration for Darwin and later evolutionists and ubiquitous textbook example of adaptive radiation.

ISBN-13:978-3-642-78965-6 e-ISBN-13:978-3-642-78963-2
DOI:10.1007/978-3-642-78963-2

Library of Congress Cataloging-in-Publication Data. Islands: biological diversity and ecosystem function/Peter M. Vitousek, Lloyd Loope, Henning Adsersen (eds.). p. cm. – (Ecological studies;v. 115) Includes bibliographical references and index. 1. Island ecology . 2. Biological diversity. I. Vitousek, Peter Morrison. II. Loope, Lloyd L. III. Adsersen, H. IV. Series. QH541.5.I8I84 1995 574.5'267--dc20 95-3817

Softcover reprint of the hardcover 1st edition 1995

Typesetting: Thomson Press (India) Ltd., Madras

SPIN: 10426981 31/3130/SPS–5 4 3 2 1 0 – Printed on acid-free paper

Foreword

The threat, and reality, of global change has caused us to examine more closely the controls on the abundances and distributions of organisms. As a result, the study of biogeography and palaeoecology is receiving renewed attention. New insights are being attained that are providing the tools to make predictions of how the world's biotic systems will be reconfigured due to changing land use, a changing atmosphere and a changing climate.

An additional element of global change is the biotic rearrangements that are occurring on Earth due to biological invasions and biotic extinctions. We have ample evidence of the massive intercontinental biotic exchanges that are occurring and of the consequences of these invasions. We also have indications of the great losses in biotic diversity that are occurring at all levels, particularly now in the tropics. For some reason more attention has been given to documenting these losses than in understanding the ecological consequences of the new biotic world in which we live.

To address this issue, SCOPE (Scientific Committee on Problems of the Environment) initiated a program to examine the "ecosystem function of biodiversity". Specifically, the program addressing two basic questions: "How do species affect ecosystem processes in the both the short and long term and in the face of global change?" and, "How is system stability and resistance affected by species diversity and how will global change affect these relationships?". These questions are species-centered, as have been most of the issues related to biodiversity. However, as the program has developed, additional attention has been given to the other changing dimensions of biodiversity, such as populations, communities, and even landscapes.

The basic approach of the SCOPE program has been to examine the functioning of the major ecosystems of the Earth to assess how they have been impacted by species losses and additions as well as by habitat fragmentation and disturbance. This review provides us with the first global assessment of the nature of the problem. The next phase of the program, to be developed by the International Geosphere-Biosphere Program (IGBP), will refine the assessment by launching a specific program to provide new information from experiments specifically designed to address the diversity/function problem rather than relying on the imperfect information derived from inadvertent "experiments" instigated by human activities or from natural environmental fluctuations.

Stanford, Spring 1995 Harold Mooney

Contents

Section C: Diversity and Ecosystem Function

List of Contributors

H. Adsersen

Department of Plant Ecology, Botanical Institute, University of Copenhagen, Øster Farimagsgade 2D, DK-1353 Copenhagen K, Denmark

T.L. Benning

Department of Biological Sciences, Stanford University, Stanford, CA 94305, USA

D. Bowden

Department of Environmental Science, Allegheny College, Meadville, PA 16335, USA

J. Cooper

Percy Fitzpatrick Institute of African Ornithology, University of Cape Town, Rondebosch 7700, South Africa

C.M. D'Antonio

Department of Integrative Biology, University of California at Berkeley, Berkeley, CA 94720, USA

T. Dudley

Pacific Institute, 1204 Preservation Park Way, Oakland, CA 94612, USA

U. Eliasson

Department of Systematic Botany, University of Göteborg, Carl Skottsbergs Gata 22, S-41319 Göteborg, Sweden

J.J. Ewel

Department of Botany, University of Florida, Gainesville, FL 32611, USA

D. Foote

Research Division, P.O. Box 52, Hawaii Natl. Park, HI 96718, USA

J.H. Fownes

Department of Agronomy and Soil Science, University of Hawaii-Manoa, Honolulu, HI 96822, USA

D. Given

Department of Horticulture, Lincoln University, P.O. Box 84, Canterbury, New Zealand

J. Hall Cushman

Department of Biology, Sonoma State University, Rohnert Park, CA 94928, USA

P. Högberg

Department of Forest Ecology, Swedish University of Agricultural Sciences, S-90783 Umea, Sweden

H.F. James

Bird Division, NHB Stop 116, Smithsonian Institution, Washington, DC 20560, USA

K. Kaneshiro

Center for Conservation Research and Training, University of Hawaii-Manoa, 3050 Maile Way, Honolulu, HI 96822, USA

T. Kendle

Department of Horticulture, Plant Sciences Laboratory, University of Reading, Whiteknights, Reading RG6 2AS, United Kingdom

L. Loope

National Biological Survey, Haleakala National Park, P.O. Box 369, Makawao, Maui, HI 96768, USA

A.W. Macdonald

WWF-South Africa, P.O. Box 456, Stellenbosch 7599, South Africa

M. Maunder

Conservation Unit, Royal Botanic Gardens, Kew, Richmond, Surrey TW9 3AB, United Kingdom

D. Mueller-Dombois

Department of Botany, University of Hawaii-Manoa, Honolulu, HI 96822, USA

J. Roughgarden

Department of Biological Sciences, Stanford University, Stanford, CA 94305-5020, USA

B. Spooner

International Institute for Environment and Development, 3 Endsleigh St., London WC1H 0DD, United Kingdom

T. Upson

Conservation Unit, Living Collections Department, Royal Botanic Gardens, Kew, Richmond, Surrey TW9 3AB, United Kingdom

P. Vitousek

Department of Biological Sciences, Stanford University, Stanford, CA 94305-5020, USA

1 Introduction–Why Focus on Islands?

PETER M. VITOUSEK[1], HENNING ADSERSEN[2], and LLOYD L. LOOPE[3]

Introduction

Research on islands has long played a fundamental part in developing our basic understanding of ecology and evolution. Both Darwin's and Wallace's insight into evolution and speciation were shaped by studies on islands (Darwin 1859; Wallace 1881); it is no coincidence that Darwin felt close to "that great fact – that mystery of mysteries – the first appearance of new beings on this earth" in the Galápagos (Darwin 1845). Even today, the Hawaiian drosophilids provide a primary standard for analyses of speciation (Carson et al. 1970; Kaneshiro, Chap. 3). More recently, ecology has been enriched by analyses of competition and character displacement (Lack 1947; Brown and Wilson 1956) and by island biogeography theory (MacArthur and Wilson 1967), which were developed and tested on islands but applied much more widely.

The reasons why islands are useful in ecological studies are straightforward. Island populations, communities, and ecosystems are self-maintaining entities with well-defined geographical limits that contain the fundamental processes, properties, and interactions of ecological systems – but they often do so in simpler ways, without the complexity of most continental systems. Moreover, the influence of particular factors that control ecological phenomena often can be understood against a relatively simple background in island systems. As examples: the evolutionary radiation of a group of plants or animals that resulted from a single founding population can be traced with a certainty rarely achieved on continents or continental islands (Eliasson, Chap. 4; Kaneshiro, Chap. 3); the influences of immigration/establishment versus extinction of species richness on islands can be analyzed, if not with certainty, then at least with a clearer focus than is possible on continents (Adsersen, Chap. 2; Roughgarden, Chap. 5); and even the individual factors controlling ecosystem biogeochemistry can be analyzed in isolation to a degree scarcely dreamt of in continental systems (Vitousek and Benning, Chap. 7).

This is not to say that the results of research on islands can be applied directly to continental systems; the greater taxonomic complexity of most continental systems, and perhaps qualitative differences between continental and oceanic island biotas (Roughgarden, Chap. 5), may defeat any simple extrapolation.

[1] Department of Biological Sciences, Stanford University, Stanford, CA 94305, USA
[2] Department of Plant Ecology, Botanical Institute, University of Copenhagen, Øster Farimagsgade 2D, DK1353 Copenhagen K, Denmark
[3] National Biological Survey, Haleakala National Park, Makawao, Maui, HI 96768, USA

Ecological Studies, Vol. 115
Vitousek et al. (eds) Islands

Rather, the understanding of ecological and evolutionary processes that is gained on islands can be used to identify important processes, support the development of theory, and test the limits of conceptual and mathematical models.

More practically, islands provide a record of humanity's interactions with biological diversity in contained areas, and of the consequences of those interactions. Because the modern epidemic of anthropogenic extinctions has hit first, and hardest, on oceanic islands, islands also provide a set of management experiments in which the consequences of different approaches to managing populations, species, and ecosystems that are now on the verge of extinction can be evaluated (MacDonald and Cooper, Chap. 15). Lessons learned from successes and failures in conserving the biological diversity of islands might guide us in developing strategies for protecting continental diversity.

This volume is designed to address the question: can the same features of island ecosystems that have proved useful in understanding of ecology and evolution, and that may help us to manage threatened populations and ecosystems, also contribute to understanding the interactions between biological diversity and ecosystem function? Island ecosystems would appear to be particularly useful for studies of diversity and ecosystem function for several reasons. First, island habitats generally contain fewer species than comparable continental habitats, so that experiments that manipulate diversity may be more manageable. Second, invasions and extinctions are widespread on islands, and they can be used to evaluate the effects of inserting and deleting species and/or functional groups (D'Antonio and Dudley, Chap. 9). Third, and most interesting, diversity varies among islands and between islands and continents in potentially useful ways.

The major reasons for variation in continental systems are climate (especially the great latitudinal pattern from the arctic to the tropics), disturbance, soil fertility, and other factors. However, the same factors that affect diversity also affect numerous ecosystem functions strongly and directly, and teasing apart the resulting interactions is not easy, indeed not often possible (Vitousek and Hooper 1993). Even in the cases where variation in diversity on continents is not tied directly to known environmental factors, it is difficult to be confident that both diversity and any related pattern in ecosystem function are *not* both controlled by another factor. (Those concerns do not apply to experiments in which diversity is deliberately manipulated as a factor, only to the many cases in which we attempt to use biogeographic pattens, dynamics, or experiments that were designed for other purposes to identify connections between diversity and ecosystem function).

In contrast, island ecosystems vary in diversity not only for the reasons described above (Loope, Chap. 10), but also, strikingly, due to rarity of successful colonization and establishment on remote oceanic islands (and also, due to island size and the time over which colonization and speciation have operated). It is possible to locate islands in which population or species-level diversity is very low, under tropical climate conditions where diversity would be substantial on a comparable area of a continent or continental island. It is also possible to locate situations in which climate and soils generally are similar on a range of islands that nevertheless differ in diversity due to their distance from a source area. For example, mangroves are a well-defined group of plants that play important roles in ecosystem function – and in the southwest Pacific the richness of mangrove taxa varies with distance from their regional center of diversity, ranging from 30 in Papua New Guinea to no native species in the Society Islands (Woodroffe 1987). Moreover, ecosystem properties and processes themselves often vary in more comprehensible ways on many islands than in most continental situations (Vitousek and Benning, Chap. 7). Consequently, both sides of the biological diversity/ecosystem function interaction may be relatively approachable on islands.

There are, however, a number of difficulties inherent to examining biological diversity and ecosystem function on islands. First, islands are not just less diverse than continental ecosystems, their disharmony may also make them different in some fundamental ways (Eliasson, Chap. 4; Roughgarden, Chap. 5). Second, few islands are unaltered by humanity, and often anthropogenic alterations have taken place on a scale even greater than that in most continental system. The disproportionate representation of island species in the lists of threatened and endangered species and of recent extinctions makes clear one consequence of those changes (James, Chap. 8; MacDonald and Cooper, Chap. 15). Consequently, to evaluate biological diversity and ecosystem function on islands, we have to deal with remnants (themselves altered) of unusual systems and/or learn to read the historical record with a precision that hitherto has been difficult to achieve (Maunder et al., Chap. 16). Nevertheless, even in their altered state, we believe that islands offer a range of systems in which diversity varies independently of other controls on ecosystem function; they therefore offer great opportunities for observation, analysis, and experiment.

How well have we made use of what islands have to offer? This volume illustrates that in understanding patterns and controls of biological diversity at all levels of biological organization from genetic to ecosystem diversity, biologists have done reasonably well (cf. Adsersen, Chap. 2; Kaneshiro, Chap. 3; Eliasson, Chap. 4; Roughgarden, Chap. 5; Foote, Chap. 6; Vitousek and Benning, Chap. 7); in using the unique attributes of islands to understand ecosystem function (Cushman, Chap. 11, Given, Chap. 12; Mueller-Dombois, Chap. 13), we are making considerable progress; in combining analyses of diversity and ecosystem function, we have a long way to go, on islands as elsewhere (Fownes, Chap. 14; Bowden, Chap. 17, Ewel and Högberg Chap. 18). We hope that this volume will help to stimulate critical research on the interactions between biological diversity and ecosystem function on oceanic islands – and elsewhere.

Acknowledgments. We thank the National Science Foundation and the Global Biodiversity Assessment for supporting the workshop that led to this volume, and Pericles Maillis and the Bahamas National Trust for hosting that workshop, and for showing us what the Bahamas can offer to island ecology.

References

Brown WL, Wilson EO (1956) Character displacement. Syst Zool 5: 49–64
Carson HL, Hardy DE, Spieth HT, Stone WS (1970) The evolutionary biology of the Hawaiian Drosophilidae. In: Hecht MK, Steere WC (eds) Essays in evolution and genetics in honor of Theodosius Dobzhansky. Appleton-Century Crofts, New York, pp 437–543
Darwin C (1845) The voyage of the Beagle. Everyman's Library, JM Dent, London, 365 pp
Darwin C (1859) On the origin of species by means of natural selection. John Murray, London, 458 pp
Lack D (1947) Darwin's finches. Cambridge University Press, Cambridge
MacArthur RH, Wilson EO (1967) The theory of island biogeography. Princeton University Press, Princeton
Vitousek PM, Hooper DU (1993) Biological diversity and terrestrial ecosystem biogeochemistry. In: Schelze E-D, Mooney HA (eds) Biodiversity and ecosystem function. Springer, Berlin Heidelberg New York, pp 3–14
Wallace AR (1881) Island life. Harper and Brothers, New York, 522 pp
Woodroffe CD (1987) Pacific island mangroves: distribution and environmental settings. Pac Sci 41: 166–185

Section A
Patterns and Levels of Diversity

2 Research on Islands: Classic, Recent, and Prospective Approaches

HENNING ADSERSEN

2.1 Introduction

In classic studies of island biota three main approaches may be identified: description of strange plants and animals as *phenomena* at the species level; examination of the *mechanisms* behind evolutionary and distributional patterns recognized at the population and community level; and quantitative and comparative evaluation of the *diversity* of entire island biota.

More recent approaches focus on how island populations or metapopulations fit into food web/trophic structure and examine how *ecosystem functioning* and *population dynamics* interact.

Strange island organisms, evolutionary mechanisms, ecosystem functioning, and population dynamics may be viewed on a background of island diversity patterns. By this means, a view of island *biodiversity* dynamics is discernible.

This *biodiversity approach* is applied on the flora of the Galápagos Islands and the Mascarene Islands, to provide a comparison of evolutionary trends, distribution patterns, and conservation needs in these two tropical, volcanic archipelagoes.

Island *ecosystem dynamics* have been studied, both from the bioenergetics view and from the aspect of ecosystem stability.

The approaches mentioned above have provided ample evidence that island ecosystems and island biodiversity are extremely *vulnerable*. This has generated a *conservation*-orientated approach, both retrospective (effect of disturbance and alien organisms) and decisive (conservation strategy).

The strong spin-off to other fields of ecology has also demonstrated the importance of island studies and underlined that island conservation is not merely a question of idiosyncratic preservation of species and habitats. Island approaches may and should provide important contributions to ecological science.

2.2 Classic Approaches

Islands fascinate and attract all imaginative and curious people (including most biologists), because they are strange units in relation to the continental situation

Department of Plant Ecology, Botanical Institute, University of Copenhagen, Øster Farimagsgade 2D, DK-1353 Copenhagen K, Denmark

Ecological Studies, Vol. 115
Vitousek et al. (eds) Islands

which is the normal situation for most of us. These strange units are furthermore normally small in size so that it is relatively easy, e. g., to comprehend the diversity of organisms that inhabit them.

2.2.1 The Dodo Approach

The oddities were the most conspicuous species to the curious spectator (a Homeric heritage?), and the study of oddities became the approach to island biology for a long time. Carlquist's *Island Biology* (Carlquist 1974), in which he identified a suite of *phenomena* peculiar to island organisms is a culmination of this approach. Among such phenomena he included loss of dispersability, unusual dimensions (gigantism or dwarfism), extreme specialization in connection with adaptive radiation, woodiness in genera or families where herbaceous growth is otherwise predominant. One may call it "the dodo approach". This approach will reveal interesting aspects of the endemics on any island. On the other hand, the majority of the endemic organisms may look quite similar to organisms in similar habitats in continental areas, and it is still undocumented that there is a larger proportion of oddities among island endemics than among plants and animals of continental origin. Elephants and giraffes are quite as odd as dodos and marine iguanas!

2.2.2 The Finch Approach

Some of the keenest spectators noted another peculiarity of islands, or at least of archipelagoes: in some cases many of the endemic species were obviously close relatives. Observations and closer studies on such cases have been extremely rewarding for development of theory. The best example is probably the Galápagos finches. Darwin's observations on these birds (which are by no means odd-looking) were essential for his formulation of the theory of natural selection (Darwin 1845), and the studies by Lack (1947), by Bowman (1961), and by Grant (1986) and his group have greatly widened our understanding of the evolutionary aspects of ecology. The approach here is to take advantage of the high degree of endemism and the low degree of complexity of islands to study the members of a clade and the *mechanisms* behind its development. It is tempting to call it the "finch approach" to island studies. The finch approach has fostered such concepts as founder effect, adaptive radiation, character displacement/release, and density compensation; but again, there are few cases where it has been considered whether the mechanisms are especially common on islands, or less common on continents.

2.2.3 The Island Biogeography Approach

Both the dodo approach and the finch approach focus on qualities of island biota. Another school focuses on the quantity and distribution of biota. In the beginning it was a purely descriptive branch of biogeography, but within the frame of what is currently recognized as island biogeography pattern recognition, causal interpretation and even experimental verification took shape. The appearance of *The Theory of Island Biogeography* (MacArthur and Wilson 1967) was a bloom of this school, but faint seedlings can be recognized already in Hooker's treatments of Darwin's plant collections from Galápagos (Hooker 1847). He both listed and counted species from the islands, and attempted to interpret the patterns exhibited by the entire flora lists for individual islands (the *florula*). In 1867 he produced what may be called the first general treatment of island biology (Williamson 1984). The further development from Palmgreen's and Arrhenius's work on islands in the Baltic (e.g., Arrhenius 1921; Palmgreen 1925) which generated the species area power curve, over Preston's derivation of a mathematical model which could generate this relationship (Preston 1962) to MacArthur and Wilson's dynamic equilibrium hypothesis is well known. Papers building on this theoretical fundament have dominated island studies since then, and are still prevalent (see, e.g., Whittaker 1992).

This development has not been a tranquil road to unanimous acceptance of the approach of the hypotheses it has generated. On the contrary, it has been a series of fierce discussions ever since the Arrhenius/Gleason (Gleason 1922) controversy and to the hot theoretical feud between the Simberloff group and the Diamond group (see Chaps. 17–20 in Strong et al. 1984) and the pro et contra contributions to the debate on the SLOSS principle in conservation (see Simberloff and Abele 1984). These discussions do not stand second in intensity to the Darwinist-anti-Darwinist feuds or the disputes on association or continuum in vegetation science, but as in these cases, the discussions have been very fruitful in terms of refinements of concepts and ideas. One thing that stands is that features on the level of island floras or florulae – one may call it the *biodiversity* level – are relevant and dynamic objects for research.

These three approaches may be termed classic approaches to island ecology. They target different levels of organization: the dodo approach focuses on species, the finch approach on clades, and island biogeography on island biota.

2.3 Recent Approaches

2.3.1 The Spider Approach

Two recent approaches target levels which are intermediate: the metapopulation studies where small populations are investigated from a population genetics view, and the food web theory, where small populations are studied in relation to their position in the trophic structure. These works do not necessarily deal with insular situations, but as the focus is on the balance between immigration and extinction for small populations, islands comprise very apt study objects. The focal questions pertain to what actually determines whether species/populations on the verge of extinction survive or not: is it a question of resilience to impact of other organisms (predation, competition) or are there minimal viable population sizes? There are works which attempt to synthesize both approaches, as Schoener's group, working on i.a. spiders in the Bahamas (Schoener 1991). I coin the term the "spider approach" for this flourishing study area.

2.3.2 The Biodiversity Approach

The biodiversity of islands is naturally restricted within narrow geographical limits. As a consequence, biodiversity on islands, by any measure, will be of a lower magnitude than in continental areas of similar ecological type. Evolving dodos and finches have not during their evolution stood up to a vast diveristy of competitors, predators, and pathogens, and spiders find themselves in small populations in simple food webs. The low biodiversity has both allowed this evolution to occur and made it very conspicuous to observe. Hence, it is actually the biodiversity pattern of islands which have made the dodo, the finch, and the spider approaches appropriate. So it would be relevant to relate such studies to the total relevant biodiversity of the islands. A first relevant question which is almost never asked is: *how common is the phenomenon or mechanism we study among the biodiversity of our island?* In other words: are we telling anecdotes or are we recognizing patterns? The second question is: *are the patterns we may consider typical island patterns, or are they found also in continental settings?*

A recent issue of *Biodiversity Newsletter* was dedicated to the biodiversity of New Caledonia. I refer to Morat (1993) and seven subsequent papers in the same issue, as prime examples of a biodiversity approach. In the following I will give other examples from the Galápagos Islands (1000–1200 km west of Ecuador's coast) and the Mascarene Islands (La Réunion, Mauritius, and Rodrigues, 800–1200 km east of Madagascar), two oceanic volcanic groups where I have been able to do studies in the field – much more in the former group than in the latter.

2.3.2.1 Endemism

The biodiversity approach to island studies has in its classic form both the notion of the number of species and the number of species particular to the island, as Hooker phrased it. Still many papers focus on endemism, as we would call it now, and with good reason, as endemism after all is easier to determine than many other effects of isolation. Williamson (1981) even proposed to let the endemism percentage define whether an island should be regarded as oceanic or continental. This aspect is easy to grasp, but there are two conditions for a good estimate of endemism percentage: good knowledge on the biodiversity on the island of the high rank taxon studied, and one must be sure that the supposed endemics do not occur somewhere else. In Fig. 2.1, I have chosen some oceanic archipelagoes for which the first criterion has been fulfilled in terms of recent publications of checklists or floras, or where databases have been available. To meet the second criterion is more delicate: it is still not possible to say that the continental areas which might harbor endemics are as well surveyed as these well-studied archipelagoes.

2.3.2.2 Isolation

With this reservation, I venture to depict the endemic percentage against an isolation measure (Fig. 2.1): distance from the center to nearest point on a continent. Immediately, a pattern emerges, and not just a trivial straight line, which would indicate that only this distance could explain endemism; but the influence of distance can be assessed, and the search for explanation for the residual variation can proceed.

Distance is obviously not the only isolation measure. We may try to approach the isolation from the biological side instead of the geographical side. This is done in Fig. 2.2, where the Galápagos and the Mascarenes are compared. The abscissa is the rough length of the dispersal element (the diaspore). This ought to have been recorded for the entire biodiversity, and then the endemic percentage within certain intervals of this scale should be assessed and depicted. A shortcut to this laborious procedure is to observe certain taxonomic groups in which the variation of diaspore magnitude is restricted.

Again a clear pattern is observed, and the pattern is remarkably uniform for the two archipelagoes. Even if they are both volcanic and have about the same distance to the continent, and the same endemic percentage among vascular

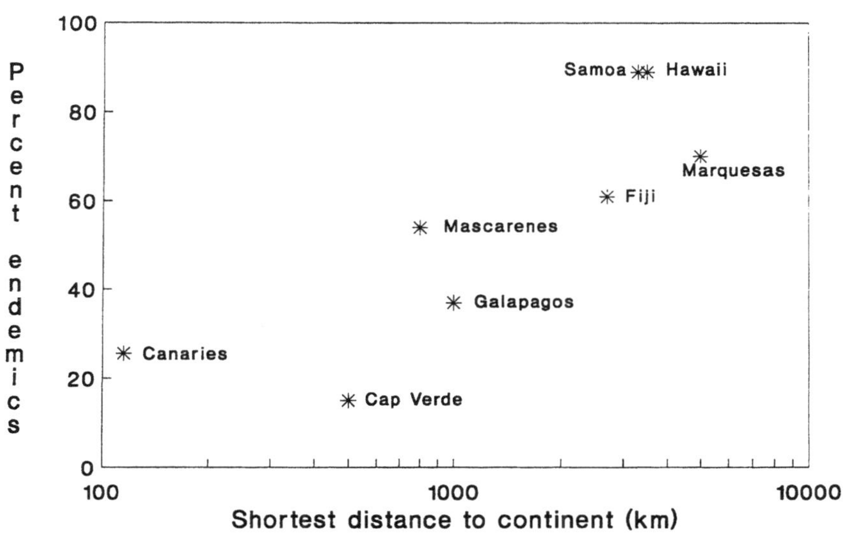

Fig. 2.1. Percentage of endemism in selected oceanic archipelagoes, against distance to closest continent (Data from Given 1992; Humphries 1977; Lawesson et al. 1987; Wagner et al. 1990; Strasberg et al., in prep.)

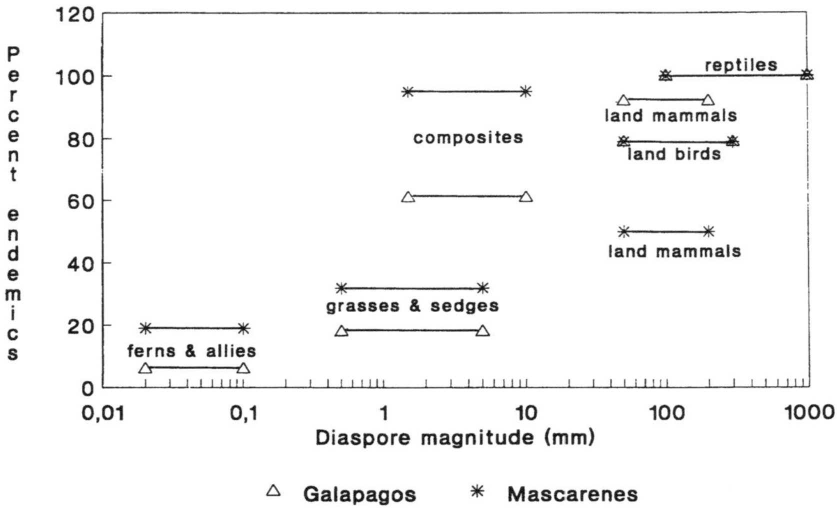

Fig. 2.2. Percentage of endemism in Galápagos and the Mascerenes in selected taxa, against diaspore magnitude. (Data from Cheke 1987; Harris 1974; Lawesson et al. 1987; Strasberg et al., in prep.)

plants, there are great climatic differences, and the donating species pools are also very different. So this striking pattern call for examination of the underlying mechanisms.

2.3.2.3 Distribution Pattern and Biodiversity

Many studies of distribution patterns within archipelagoes state certain groups of species as typical for a certain distribution type (e.g. Greutner 1979). Such observations are interesting, but as examples they ought to be seen in relation to the entire biodiversity. One may ask the question: do we find more or fewer plants with such particular distribution than should be expected with the given biodiversity in the archipelago? A less complex, but related question is: *do the islands in the archipelago pairwise share less or more species than expected?* Connor and Simberloff with their null hypothesis approach were the first to try to solve the problem (Connor and Simberloff 1978; Simberloff 1978); later, an explicit formula was suggested (Adsersen 1988, 1990). Recently, this approach has been used for a variety of taxa within the Canary Islands by Fernández-Palacios and Andersson (1993). One advantage with this approach is that one uses the information both on the number of species on each island, and on their distribution. They were able by this analysis to sort out which Canany Islands taxa were not randomly distributed among the islands, and could suggest explanations.

2.3.2.4 "Dodos" and Biodiversity

With a knowledge of island biodiversity we may ask the relevant question: how frequent is the evolution of some of the features identified as typical for islands under the dodo approach? One could question whether insular woodines is a common feature or not. Within Galápagos, insular woodiness can be found in the four endemic genera of composites: *Scalesia*, *Darwiniothamnus*, *Macraea*, and *Lecocarpus*. *Alternathera* (Amaranthaceae) provide another example, and with some hesitation I include also the arboreal or shrubby succulents of *Calandrinia* (Portucaceae) and *Opuntia* (Cactaceae). These 7 genera account for a total of 51 endemic species, or 23% of the endemics. Is this so many that one would call it a characteristic feaure? Another question is whether this is the most relevant calculation. Would it not be more relevant to ask how many of the immigrated species gave rise to insular woodiness? *Scalesia* comprises the largest genus with respect to number of endemics. *Scalesia*'s 21 taxa are derived from one immigration event. *Alternathera*'s endemics stem from three different immigration events (Porter 1983, 1984), with only one of them (the *Alternathera filifolia* complex) demonstrating insular woodiness. The arboreal *Opuntias* descend from two profusely scrambling immigrants (Porter 1984). Hence, insular woodiness is occurring in only seven of the evolutionary lines in Galápagos. The total number of evolutionary lines in Galápagos amounts to 104 (7%). If we turn to the Mascarenes, most of the speciation took place in families where woodiness was predominant, so even if there are many woody plants among the endemics in the Asteraceae, the analogous percentages are lower, and may not be different from figures from continental situations. The relevant figures are shown in Table 2.1.

In the endemic flora of the Mascarenes, another feature is much more striking than insular woodiness: the phenomenon of heterophylly. Friedmann and Cadet (1976) described how 45 endemic species showed a tendency to have smaller or more finely dissected leaves in a juvenile stage than as adults. This accounts for 4% of the entire flora or 7% of the endemic flora, and it occurs in no less than 17% of the endemic lines. Is this an island feature? Friedmann and Cadet give no examples of a similar importance of heterophylly on other islands, so the comparison must be made with continental areas, and this is only possible on the biodiversity level.

Table 2.1. "Dodo approach" features: The number and percentage of the total flora that display insular woodiness and heterophylly. (Data from Friedmann and Cadet 1976; Adsersen 1991; Strasberg et al., in prep.)

	Endemics		Insular woodiness				Heterophylly	
	Galá-pagos	Masca-renes	Galápagos no.	Galápagos %	Mascarenes no.	Mascarenes %	Mascarenes no.	Mascarenes %
ssp. and subspp.	219	641	51	23	50[a]	8[a]	45	7
Lines	104	211	7	7	6[a]	3[a]	35	17[a]

[a]Approximate figures.

2.3.2.5 "Finches" and Biodiversity

In the finch approach, the main focus is on radiative groups, and it has often been stated that radiation is a typical island feature. The classical example is that of the Darwin finches from Galápagos: 13 endemic species have evolved from one common ancestor. How does this look in the biodiversity background? Galápagos has 28 indigenous land birds; 22 of these are endemics, yielding an endemic percentage of 79% (Harris 1974). Apart from the finches, the Galápagos mockingbirds show radiation with four endemic species. Thus, 17 of 22 endemics (77%) are due to radiation, if we count resulting species. We may, however, count the ancestral immigrants instead of the present biodiversity. Then we have the result that of the 13 immigrants that still have descendants in Galápagos, 46% did not change, 39% changed to another species by unidirectional evolution, and 15% underwent radiative evolution. In the Galápagos avifauna, it is easy to define radiation – it is more difficult in other islands or taxa. In Table 2.2, radiation is defined as the condition with more than one endemic species in a genus. It is a very inclusive definition, but it

Table 2.2. "Finch approach" features: the number and percentage of indigenous species and lines that are endemic or have radiated[a]. (Data from Adsersen 1991; Cheke 1987; Harris 1974; Strasberg et al., in prep.)

	Indigenous	Natives		Endemics		Radiatives	
	spp./lines	spp.	Indigenous spp. (%)	spp.	Indigenous spp. (%)	spp.	Indigenous spp. (%)
Galápagos land birds							
Species	28	6	21	22	79	17	61
Lines	13	6	46	7	54	2	15
Mascarene land birds							
Species	56	10	18	46	82	28	58
Lines	39	10	29	29	74	12	31
Galápagos plants							
Species	569	350	62	219	38	151	27
Lines	454	350	77	104	23	36	8
Mascarene plants[b]							
Species	1131	490	43	641	57	330	29
Lines	701	490	70	211	30	100	14

[a]The data for lines in Galápagos represent phylogenetic lines according to taxonomists's view. This has not been possible to establish for Mascarene organisms, instead the number of genera with more than one endemic has been counted as a radiative line. There are two sources of errors in this: more than one endemic line may arise within one genus (as in *Alternathera* from Galápagos, three lines), or evolution may proceed out of the level of species, so some endemic genera may belong to the same line (like the finches in Galápagos, six endemic genera in one line). The two sources of errors counteract each other.

[b]Preliminary figures, based on the families that have until now been treated in recent floras or flora drafts.

allows comparison between archipelagoes and taxa where exact phylogenies are not sufficiently known. The indigenous avifauna of the Mascarenes is treated by Cheke (1987). Apart from the sad fact that 61% of the species are extinct now, and that many are only imperfectly known from incomplete subfossils or even anecdotes, it is possible to set up a data set to compare with Galápagos. There is a striking accordance on the species level for the two groups, but if one considers the data for lines, it shows that there are differences. Apparently, radiation, using the inclusive definition, is more common in the Mascarenes. If only genera with more than three endemics are considered radiative, only four lines qualify, and no line in the Mascarenes counts more than three endemics – one on each of the three major islands. Table 2. 2 also gives the figures and the corresponding figures for plants of the Galápagos and the Mascarenes. Have we overestimated the frequency of radiation?

Another question is whether one should speak of *adaptive* radiation. To me, adaptive radiation means that one has several successful species evolved sympatrically into different niches where they may acquire evolutionary stability. The Darwin finches are an example of this; but if we look at plants, the radiation more often leads to "weak" species which are very local and obviously have to evolve further in order to avoid extinction. I have tried (Adsersen 1991) to elucidate this from a biodiversity approach, and the result was that *adaptive* radiation is far less frequent than the other type of radiation, for which I suggest the term *fugitive* radiation. Actually, 58% of the plant species arising from radiative evolution in Galápagos are not very successful in terms of dominance or abundance, of large stands, or large distribution. They are probably transient species.

2.3.2.6 "Spiders" and Biodiversity

The biodiversity approach allows one to compare frequencies of island features, within and between islands, and between islands and continents. Useful comparisons can be made of features of the endemic part of the flora and the native part (defined as the indigenous, nonendemic part of the flora). We have performed an analysis of a food web trait in this manner: the ability of plants to produce cyanogenic compounds may be assumed to be a herbivore-repelling mechanism. We examined this ability within the endemic and the native contingent of the Galápagos flora (Adsersen and Adsersen 1993), and it proved that there are more cyanogenic plants among the endemic than among the native plants (Table 2.3). Is this caused by selection in the particular herbivory regime on islands where mammal herbivores are absent?

2.3.2.7 Conservation and Biodiversity

A similar approach was taken to evaluate the state and threats to the rare plants of Galápagos (Adsersen 1989). It showed that of the 226 endemic species and subspecies considered, 69 could be called rare (30%), whereas of the 378 native species and subspecies recognized, 75 were rare (20%). Rareness need not imply

Table 2.3. "Spider approach" features: cyanogenesis in plants from Galápagos[a]

	Natives				Endemics			
	examined	cyanogenic	slightly cyanogenic	non cyanogenic	examined	cyanogenic	slightly cyanogenic	non cyanogenic
No. of spp.	237	32	28	177	120	20	26	74
Percentage	100	13.5	11.8	74.7	100	16.7	21.6	61.7

[a] Cyanogenetic plants produce more than 10 mg HCN/kg fr. wt. after incubation with glycolytic enzyme, slightly cyanogenic produce between 2.5 mg/kg fr. wt. and 10 mg/kg fr. wt. For details on methods and interpretations, see Adsersen and Adsersen (1993).

vulnerability, but a closer look revealed that for 67% of the rare endemics, abundance or frequency decrease could be demonstrated, whereas the corresponding figure for the natives is 35%. It also proved that introduced animals threaten the existence or performance somewhere in the archipelago for 61% of the rare endemics, whereas the corresponding figure is 20% for the rare natives. Introduced plants exert threats to 29% of the rare endemics, and to 11% of the natives.

The figures are summarized in Table 2.4. In rough terms: rareness is 1.5 times as frequent among endemics as among natives. Decrease in abundance hits twice as many rare endemics as rare natives, and rare endemics are three times more vulnerable to introduced animals and plants than are rare natives.

2.3.3 The Ecosystem Functions Approach

The biodiversity studies clearly provide a holistic approach. Holistic approaches have been in the background for many years, since the first international

Table 2.4. Conservation features: Status and threats to plants in Galápagos. (Data from Adsersen 1989)

	Native plants			Endemic Plants		
	no.	Total spp. (%)	Rare spp. (%)	no.	Total spp. (%)	Rare spp. (%)
Total spp. and subspp. considered	378	100		226	100	
Rare spp. and subspp. considered	75	19.8	100	69	30.5	100
In serious decrease	20	5.3	26.7	40	17.7	58.0
Threatened by alien animals	15	4.0	20.0	42	18.6	60.9
Threatened by alien plants	8	2.1	10.6	20	8.8	29.0

ecological congress with the title: Unifying Concepts in Ecology (van Dobben and Lowe-McConnell 1975). Maybe the only unifying concept at that time was that nothing conclusively could be expressed on the holistic relationships among ecosystem parameters like productivity, resource availability, diversity, and stability. This could be one of the reasons why ecosystem dynamics studies have been uncommon in island studies, with some exceptions: Wright (1983) suggested that biodiversity on islands be related to such ecosystem parameters as potential productivity and evaporation (Wright 1983). I have reserved another contribution to be mentioned last but not least: Mueller-Dombois et al. (1981): *Island Ecosystems*, which is actually a monograph on the ecology of Hawaii. Most of the above-mentioned approaches are represented in this book, and they are nicely tide up in a synthesis.

2.3.4 The Conservation Approach

Almost every aspect of the above approach has given evidence to what we knew: island biodiversity and ecosystems are extremely vulnerable. A large amount of work has concentrated on why islands are vulnerable, and how they could be protected. We have learnt a lot of what happens when alien organisms (including man) intrude into islands: the dodo, several populations of Darwin finches, and I am sure some spiders have already become extinct in the definitive sense of the word, and ecosystem degradation seems to be the rule rather than the exception. Land use for agriculture, infrastructure, urbanization, etc. is an ubiquitous trivial menace, but even island ecosystems under conservation protection seems to retain an innate vulnerability. It is tempting to set up a metaphor like this: breakdown of biodiversity isolation causes island ecosystem disturbance. So in order to preserve island ecosystems, we have to watch that island biodiversity is not contaminated. This may be one of the main differences between continental and island situations: the main worry in continental reserves may be ecosystem contamination in terms of chemical pollutants, whereas island reserves suffer from biodiversity contamination in terms of alien organisms.

2.4 Recapitulation

2.4.1 Islands as Model Ecosystems

As demonstrated by the examples above, biodiversity of islands can be efficiently assessed, and ecological features of their biota may be quantitatively expressed in both absolute and relative terms. This is one reason why islands are excellent objects for biodiversity studies. Another reason is that the extent to which biodiversity of islands has been disturbed is overwhelming, in terms of both extinction of indigenous organisms and introduction of aliens. So island systems provide a very wide range of biodiversity variation.

Ecosystem *structure* may be conceived as the compartmentalization of the organisms present, in terms of species composition and food web position. Hence the variation of biodiversity of islands rends a series of ecosystems which vary greatly in structure.

Ecosystem *functioning* is the result of the biological turnover of the physico-chemical part of the environment – climate, parent material, and topography in classical terms of Jenny (1980) and Major (1951). There is no reason to believe that the variation of weather, soil, and relief should be greater in insular situations than on continents (rather the opposite), so it is not surprising that Mueller-Dombois (1981) concluded that island ecosystem functioning did not differ convincingly from that in continental situations.

Ecosystem *dynamics* describes how structure and function changes over time scales longer than the diurnal and seasonal oscillations.

For most islands we have rather precise ideas of their age and the history of their formation. The time scales involved vary from geological to generational time. Furthermore, for most islands we have good estimates or precise records on when man arrived and initiated human impact.

Hence islands contribute study areas where there is a great variability in ecosystem structure caused by differences in biodiversity, where the well-studied ecosystem functions are in action, and where we may have good knowledge of starting points (island ages) and milestone events (arrival of humans, disappearance or advent of key species). What more can we demand of model systems for study?

2.4.2 Isolates that are Not Islands

True islands are landmasses surrounded by water. This geographical definition is, however, not what makes islands interesting in an ecological context. The key point is that their biotas are isolated within a restricted area, so that contact with conspecifics or species which interact ecologically requires that dispersal barriers are overcome. With this ecological definition, many habitats or landscapes become isolates, for which many of the considerations above hold true. The school of island biogeography has been long aware of this, and has studied probably as many habitat isolates as true islands. Examples are lakes, mountains, caves, wood lots, and even individual plants.

There are, however, for most habitat islands, features which separate them from true islands. In classical island biogeography one considers *insular organisms*, which are partially separated by *hostile barriers* from *continental species pools*. These features pertain the relation of isolate biota to biota in "the rest of the world", so they are biodiversity features and therefore important to bear in mind if biodiversity/ecosystem relations are to be studied.

True islands are separated from continents of high diversity by a sea of low biodiversity, and there is a large pool of potential immigrants. Isolated mountains generally emerge above landscapes which contain many terrestrial species, so the dispersal barrier is here a "sea" of high biodiversity. For lakes there is a third situation: their biota experience the terrestrial surroundings as barriers of low biodiversity (unless the organism has an amphibian life cycle as many insects), and there is hardly any continental species pool. These traits do not exclude habitat isolates as model systems in our context, rather the opposite: it is possible to select systems with any wanted magnitude of biodiversity in the barrier and in the species pool.

2.5 Prospective Approaches

The remarks above address island ecosystems from a holistic view. There is generally a need for more holistic views of island ecology. There is here a lot of desk work to do to compile existing knowledge, and a lot of field work to do to fill known or unknown gaps.

To be more specific, I would like to see the following approaches:

- Research on dodos, finches, and spiders in relation to total biodiversity of islands.
- Research on the role of dodos, finches, and spiders in island ecosystem dynamics.
- Comparative research on diversity of islands and search for common features: an invasibility approach.
- Comparative research on alien organisms in relation to island ecosystems: an invasivity approach.

These are pleas for research works. During this past year I have had the luck to visit four oceanic archipelagoes: the Mascarenes, the Galápagos, the Canaries, and the Bahamas. I have noticed the same invasive plants everywhere, and practical conservationists speak about the same problems; but direct contact between conservationists in these four archipelagoes is literally nonexistent. A workshop like this one brings together scientist colleagues for fruitful discussions and exchange of experience. I believe there is a great need of similar opportunities for conservation practitioners as well. As much as we researchers appreciate islands as scientific playgrounds, it becomes evident to us that we owe practical conservationists our best support to keep island biodiversity and ecosystems existing, vital, dynamic, and enchanting.

The approaches to island ecology have had strong spin-offs to other fields of science. Evidence from island studies has been crucial for current scientific theories within evolution, plate tectonics, climatology, cultural anthropology, cladistic taxonomy, and conservation. Island conservation clearly involves much more than mere idiosyncratic preservation of species and habitats. There is every reason to believe that island approaches will continue to provide important contributions to ecological science. Islands may reveal important relationships between biodiversity and ecosystem functioning.

Acknowledgments. I thank the Galápagos National Park Service and the Charles Darwin Research Station for research permission and constant interest and support. The field work was made possible through grants from UNESCO, DANIDA, The Danish Natural Science Foundation, Julie von Müllen's Foundation, and WWF International.

References

Adsersen A, Adsersen H (1993) Cyanogenic plants in the Galápagos Islands: ecological and evolutionary aspects. Oikos 67: 511–520

Adsersen H (1988) Null hypotheses and species composition in the Galápagos Islands. In: During HJ, Werger MJA, Willems JH (eds) Diversity and pattern in plant communities. SPB Academic Publishing, The Hague, pp 37–46

Adsersen H (1989) The rare plants of the Galápagos Islands and their conservation. Biol Conserv 47: 49–77

Adsersen H (1990) Intra-archipelago distribution patterns of vascular plants in Galápagos. In: Lawesson JE, Hamann O, Rogers G, Reck G, Ochoa H (eds) Botanical research and management in Galápagos. Monogr Syst Bot Mo Bot Gard 32: 67–78

Adsersen H (1991) Evolution, extinction and conservation. Examples from the Galápagos flora. Evol Trends Plants 5: 9–18

Arrhenius O (1921) Species and area. J Ecol 9: 95–99

Bowman RI (1961) Morphological differentiation and adaptation in the Galápagos finches. Univ Calif Publ Zool 58:1–302

Carlquist S (1974) Island biology. Columbia University Press, New York

Cheke AS (1987) An ecological history of the Mascarene Islands, with particular reference to extinctions and introductions of land vertebrates. In: Diamond AW (ed) Studies of Mascarene Island birds. Cambridge University Press, Cambridge, pp 5–89

Connor EF, Simberloff D (1978) Species number and compositional similarity of the Galápagos flora and avifauna. Ecol Monogr 48: 219–248

Darwin C (1845) The voyage of the Beagle. Everyman's Library, JM Dent London (Reprint from 1967) pp 365

Fernández-Palacios JM, Andersson C (1993) Species composition and within archipelago co-occurrence patterns in the Canary Islands. Ecography 16: 31–36

Friedmann F, Cadet T (1976) Observations sur l'hétérophyllie dans les îles Mascareignes. Adansonia Sér 2, 15: 423–440

Given DR (1992) An overview of the terrestrial biodiversity of Pacific Islands. Mimeographed report to the South Pacific Regional Environment Programme, Apia, Western Samoa, pp 1–34

Gleason HA (1922) On the relationship between species and area. Ecology 3: 158–162

Grant PR (1986) Ecology and evolution of Darwin's finches. Princeton University Press, Princeton

Greutner W (1979) The origins and evolution of island floras as exemplified by the Aegean Archipelago. In: Bramwell D (ed) Plants and islands. Academic Press, London, pp 87–106

Harris M (1974) A field guide to the birds of Galápagos. Collins, London

Hooker JD (1847) On the vegetation of the Galápagos Archipelago, as compared with that of some other tropical islands and of the Continent of America. Trans Linn Soc Lond 20: 235–262

Humphries CJ (1977) Endemism and evolution in Macaronesia. In: Bramwell D (ed) Plants and islands. Academic Press, London, pp 171–199

Jenny H (1980) The soil resource. Origin and behavior. Ecological Studies 37, Springer, Berlin Heidelberg New York

Lack D (1947) Darwin's finches. Cambridge University Press, Cambridge

Lawesson JE, Adsersen H, Bentley P (1987) An updated and annotated check list of the vascular plants of the Galápagos Islands. Rep Bot Inst Univ Aarhus 16: 1–74

MacArthur J, Wilson EO (1967) The theory of island biogeography. Princeton University Press, Princeton

Major J (1951) A functional, factorial approach to plant ecology. Ecology 32: 392–412
Morat P (ed) (1993) New Caledonia: a case study in biodiversity. Biodivers Lett 1, No 3/4 (Spec Issue): 69–129
Mueller-Dombois D (1981) Island ecosystems: What is unique about their ecology. In: Mueller-Dombois D, Bridges KW, Carson HL (eds) Island ecosystems. Biological organization in selected Hawaiian communities. Hutchinson Ross, Stroudsburg, pp 485–501
Mueller-Dombois D, Bridges KW, Carson HL (eds) (1981) Island ecosystems. Biological organization in selected Hawaiian communities. Hutchinson Ross, Stroudsburg
Palmgreen A (1925) Die Artenzahl als Pflanzengeographischer Charakter. Fennia 46, 2, pp 1–144
Porter DM (1983) Vascular plants of the Galápagos: origins and dispersal. In: Bowman RI, Berson M, Leviton AE (eds) Patterns of evolution in Galápagos organisms. Pacific Division, AAAS, San Francisco, pp 33–96
Porter DM (1984) Endemism and evolution in terrestrial plants. In: Perry R (ed) Key environments. Galápagos. Pergamon Press, Oxford, pp 85–99
Preston FW (1962) The canonical distribution of commonness and rarity. Part I. Ecology 43: 185–215; Part II. ibid 43: 410–432
Schoener TW (1991) Extinction and the nature of the metapopulation: a case system. Acta Oecol 12: 53–75
Simberloff D (1978) Using island biogeographic distributions to determine if colonization is stochastic. Am Natur 112: 713–726
Simberloff D, Abele LG (1984) Conservation and obfuscation: subdivision of reserves. Oikos 42: 399–401
Strasberg D, Thebaud C, Dupont J, Adsersen H Database on the flora of the Mascarenes. (in prep) Preprints available on request to H. Adsersen
Strong DR, Simberloff D, Abele LG, Thistle AB (eds) (1984) Ecological communities. Conceptual issues and the evidence. Princeton University Press, Princeton
Van Dobben WH, Lowe-McConnell RH (eds) (1975) Unifying concepts in ecology. Junk, The Hague
Wagner WL, Herbst DR, Sohmer SH (eds) (1990) Manual of the flowering plants of Hawai'i. University of Hawai'i Press, Honolulu
Whittaker RJ (1992) Guest editorial: stochasticism and determinism in island ecology. J Biogeogr 19: 587–591
Wiggins IL, Porter DM (eds) (1971) Flora of the Galápagos Islands. Stanford University Press, Stanford
Williamson M (1981) Island populations. Oxford University Press, Oxford
Williamson M (1984) Sir Joseph Hooker's lecture on insular floras. Biol J Linn Soc 22: 55–77
Wright DH (1983) Species-energy theory: an extension of species-area theory. Oikos 41: 496–506

3 Evolution, Speciation, and the Genetic Structure of Island Populations

K.Y. KANESHIRO

3.1 Introduction

Islands have long been recognized as the best places on earth for investigating evolutionary processes. Darwin (1859) formulated his ideas about natural selection and evolution during his field studies of the organisms on the Galapagos Islands. Life forms on islands epitomize nature's creativity as evidenced by the explosive radiation and the tremendous numbers of species in many groups of organisms that have evolved on islands. On the other hand, island ecosystems also epitomize nature's vulnerability as evidenced by the high rate of extinction. Many of these island ecosystems are faced with an *extinction crisis* primarily due to the ecological fragility of islands. Thus, islands are extremely important not only for the field of Evolutionary Biology but also for the field of Conservation Biology, and the plants and animals that have evolved on islands have given scientists the opportunity to research on both these areas of biology.

There is a profound association between applied conservation biology and the study of island ecosystems. It is generally recognized by conservation scientists that even continental natural habitats or ecosystems will eventually become fragmented into smaller isolates of what were formerly larger continuous natural habitats (Wilcox 1980). Thus, ecological studies of the effect of landscape fragmentation on biological diversity will play an increasingly important role in land-use decisions and the development of management practices that conserve biological diversity while meeting the complex needs of a modern society. For this reason, ecologists have come to realize the importance of investigating island ecosystems for understanding conservation problems worldwide.

3.2 The Biota of the Hawaiian Islands

The Hawaiian Archipelago is considered by many scientists to the be *world's most outstanding living laboratory for the study of evolution and island biology.* The Hawaiian islands are home to more endemic species than any place of similar size on earth. The only tropical rainforests in the United States are found there. There are spectacular examples of adaptive radiation in the plants of Hawaii such as can

Center for Conservation Research and Training, University of Hawaii, 3050 Maile Way, Gilmore 310, Honolulu, Hawaii 96822, USA

Ecological Studies, Vol. 115
Vitousek et al. (eds) Islands

be seen in the silversword species group (Carr et al. 1989) and the Hawaiian *Bidens* (Ganders 1989). There are the amazing bird species, such as the Hawaiian honeycreepers, the Hawaiian goose (the Nene), and a number of yet to be described fossil species which are now extinct. It is estimated that there may have been as many as 150 species in the native avifauna prior to the arrival of the Polynesians (Olson and James 1982). Then, there are the insects of Hawaii; it is estimated that there may be as many 10000 species of insects in the Hawaiian fauna (Howarth and Mull 1992). Indeed, the plants and animals of Hawaii are magnificient treasures of these island ecosystems and they afford biologists an opportunity to investigate every aspect of evolutionary biology.

I have been involved in a project to investigate the evolutionary biology of the endemic insects in the family Drosophilidae of the Hawaiian Islands. The Hawaiian *Drosophila* Project has been a team effort with more than 70 senior scientists from all over the world, who have come to Hawaii for the past 30 years to investigate every aspect of the biology of these species; everything from the morphology and taxonomy, its ecology and behavior, genetics and physiology, to molecular biology and DNA sequencing. Thus far, we have been able to name and describe more than 500 species that are unique to the Hawaiian Islands and found nowhere else in the world. We have in the collection another 250–300 more new species that are unnamed and undescribed, and we continue to discover new species as we continue with our field studies. My conservative estimate is that there will be over 1000 species in this family by the time the taxonomic treatment is completed. Considering the extremely small land mass and the geologically young land mass, we have more than a quarter of the total number of species in this group found in the rest of the world.

The question is HOW COME? What is it about the Hawaiian Islands that have given rise to this tremendous flowering of species in such a small land mass? What is it about the biology, the genetics, the behavior of these species that promotes rapid speciation?

3.3 Geological Features of the Hawaiian Islands

First and foremost is the fact that the Hawaiian Archipelago is the most isolated island system on earth situated in the middle of the vast Pacific Basin. The nearest continental land mass is about 4000 km away, and the nearest island group about 1600 km. Chance migrants from the nearest land mass are a rare event at best and it is a reasonable assumption that gene flow between the ancestral and the incipient population is nonexistent. Several lines of evidence suggest that the entire fauna of Drosophilidae in Hawaii, the 800 or perhaps 1000 species, all evolved from a single (or, at the most, two) ancestral founder that arrived in these islands a few millions of years ago (Carson et al. 1970; Throckmorton 1975; Hardy and Kaneshiro 1981).

Another important feature of the Hawaiian Islands is that each island was formed in chronological sequence as the Pacific Tectonic Plate moved in a northwesterly direction over a stationary hot spot in the earth's mantle. Thus, each island is progressively younger in a southeasterly direction, with Kauai the oldest (5–6 million years) and Hawaii (less than 1 million years) the youngest of the present high islands (Dalrymple et al. 1973). In our early studies to catalog and understand the distribution patterns of the drosophilids in Hawaii, we discovered that each island had its own fauna of endemic drosophilid; in other words, those species on the Island Kauai were found only on Kauai, those on Oahu were found only on Oahu, those on the Island of Molokai were found only on Molokai (although for many species groups, the Islands of Maui, Molokai, and Lanai are considered a single biological island and collectively referred to as Maui Nui; i.e., certain species may be found on more than one of the Maui Nui Islands), and so on. Ninety-eight percent of the species were single-island endemics and only a few species were found on more than one island. In most cases, phylogenetic studies indicated that the most ancestral species were found on the older islands although there are some exceptions.

Another interesting feature about Hawaii is that even within individual islands, special geological and topographical features presented a situation where populations can become isolated and thereby differentiate into separate genetic entities. For example, on the older islands, erosion formed deep valleys and geographic barriers to gene exchange between adjacent populations at least for species with limited dispersal capabilities. On the younger islands, volcanic activity can fragment forested areas into a mosaic pattern of isolated vegetation called kipukas. The inhabitants of such patches may be genetically isolated from other populations found in a patch of forest perhaps only a couple of hundred meters away. When we began to look at the genetic differences of the drosophilid populations that were trapped in these separate kipukas, we found significant differences even though the populations were isolated only about 100 years ago.

Another significant feature of the Hawaiian Islands is that, within relatively short distances, we can find extremely different climatological characteristics. The prevailing northeast trade winds carry warm, moist air toward the islands, resulting in high rainfall on the windward (northeastern), and dry, desert-like conditions on the leeward (southwestern) sides of the islands. Wet rainforest habitats with as much as 6350 mm of rainfall per year can be found on the windward side of mountains and highlands of the islands, while dry-land forests, where there may be as little as 635 mm of rainfall for the entire year, are found on the leeward side on the islands. These conditions create a situation where the plants and animals of Hawaii are presented with a tremendous diversity of habitats within relatively short distances from one another.

3.4 Natural Selection Versus Sexual Selection

The special extrinsic features of the Hawaiian biota have played a significant role in the evolutionary history of the endemic drosophilid fauna of Hawaii. However, despite what appears to be a terrific example of adaptive radiation, it turns out that natural selection is not the most important force during the initial stages of species formation. Rather, what we have found is that it is shifts in the complex mating system, or changes within the sexual environment, if you will, that have been most important during the early stages of species formation. While the forces of natural selection are still considered to be important in directing the course of evolution in a population, adaptive shifts in response to the external environment occur over a long period of evolutionary time. On the other hand, I suggest that adjustments to the mating system, together with other pleiotropic effects in the genetic makeup of the population, occur during the most dynamic stages of the speciation process (Kaneshiro 1980, 1983, 1988, 1989; Kaneshiro and Boake 1987; Kaneshiro and Giddings 1987).

3.5 Classical Sexual Selection Theory

While Darwin (1871) first proposed the concept of sexual selection more than a century ago, it has only been within the past 15 years or so that evolutionary biologists have paid more attention to the significance of sexual selection on speciation models (see West-Eberhard 1983, for a review). Nearly 65 years ago, Fisher (1930) examined the genetic consequences of Darwin's ideas and more recently, O'Donald (1977, 1980), Lande (1981, 1982), Kirkpatrick (1982), and several others have applied a mathematical treatment to Fisher's interpretations. One of the most significant ideas that was developed by Fisher (1930) is that sexual selection via female choice results in the rapid coevolution of a male phenotype that confers mating advantage and the female's preference for that trait, i.e., "Fisher's runaway selection model". The result is a runaway selection toward greater and greater exaggeration of the male trait until natural selection acts to counter the runaway system, favoring an optimum phenotype for the particular environment in which the population lives. Thus, it is assumed that two factors act to counterbalance the runaway process of sexual selection; on the one hand, female preference for a certain male trait tends to select for extreme forms of that character, while natural selection exerts its forces to maintain the optimum phenotype for the population to survive in its environment. Mayr (1972) stated that "natural selection will surely come into play as soon as sexual selection leads to the production of excesses that significantly lower the fitness of the species...".

3.6 The Differential Selection Model

Results of mating studies of Hawaiian *Drosophila* (Carson 1986; Kaneshiro 1987, 1989) indicate that as few as 30% of the males may accomplish more than 50% of the matings in the population, while as many as 30% of the males may not be successful at all in securing a mate. Similarly, it was found that there are females that are highly discriminant and those that are not so discriminant in mate choice. As many as 30% of the females may not mate even if allowed to be courted by several males taken randomly from the population. Presumably, these females that are highly discriminant in mate choice may never mate unless they encountered and are courted by males of high courtship ability. The results of these studies indicate that there is indeed a wide range of mating types segregating in both sexes of the population, contrary what one might have expected from the classical sexual selection models, which suggest that there should be a decrease in variability because of the coevolution of male traits and female preference for sexually selected traits.

An alternative model of sexual selection was developed to account for the maintenance of variability in the mating behavioral phenotypes in the two sexes generation after generation even in laboratory cultures which may have been exposed to bottleneck situations (Kaneshiro 1989). I suggested that under normal conditions, when the population is large, the most likely matings are between males that are highly successful in mating ability (the studs), and females that are less discriminant in mate choice. The genetic correlation between these two behavioral phenotypes in the two sexes maintains the range of mating types in the two sexes in the subsequent generations. In this model, then, there is differential selection for opposite ends of the mating distribution and therefore, sexual selection itself serves as the stablizing force in maintaining a balanced polymorphism in the mating system of the population without having to invoke the forces of natural selection to counterbalance a runaway situation.

Under conditions when the population is drastically reduced, there would be even greater cost to females that are highly discriminant in mate choice. These females may never encounter males that are able to satisfy their courtship requirements, and during periods when the population is small, there would be even stronger selection for less discriminant females. If the condition persists over several generations, there would be a shift in the distribution of mating types in the population until there is a significant increase in frequency of less discriminant females in the population (Kaneshiro 1989). Such a shift in the mating distribution may be accompanied by a shift in the gene frequencies in the population, resulting in the destabilization of the coadapted genetic system that had evolved in the population. Under these conditions, novel genetic recombinants may be generated, some of which may be better adapted to the environmental stress that led to population reduction. Thus, the dynamics of the sexual selection system can play an extremely important role in the speciation process and the evolutionary history of a population.

3.7 Small Populations and Founder Event Speciation

The most likely mode of speciation in the Hawaiian Drosophilidae is what evolutionary biologists call founder event speciation (Mayr 1942; Carson 1971). The most probable scenario is that a single fertilized female is blown to an adjacent island and establishes a new colony there. For example, a fertilized female is blown from Kauai to Oahu and lands in a new habitat where the female can lay a few eggs and produce a few offspring. For a number of generations, during the initial stages of colonization when the population size is small, there will be strong selection for females that are less choosey in mate choice. Those females that are too choosey may never encounter males that are able to satisty their courtship requirements and therefore for a few generations there will be a shift in the distribution of mating types and there will be an increase in females that are less discriminant. As suggested above, these conditions may result in a significant shift in gene frequencies in the population with subsequent destabilization of the co-adapted genetic system. Genetic recombinants which might be better adapted to the new environment may be generated and strongly selected especially if these genotypes are linked with the genotypes of less discriminant females. Thus, at least during the initial stages of colonization immediately following the founder event, the dynamics of the sexual selection system may be playing a significant role in generating a genetic milieu conducive to the formation of a new species.

3.8 The Biology of Small Populations

Clearly, an understanding of the biology of small populations is crucial for understanding evolutionary processes. However, it is also important for addressing conservation issues, since threatened and endangered species are faced with extinction primarily as a result of drastically reduced population size. No matter what biotic or abiotic factors were involved in population reduction, whether it is due to catastropic environmental events such as severe drought conditions or high-intensity tropical storms such as hurricanes or typhoons, or whether it is due to a fatal disease that decimates a large proportion of the population, recovery of the population is dependent on how the population can overcome the stress condition during small population size. By applying the differential sexual selection model presented above for species formation, we find a similar situation here that during periods when the population is severely reduced, the shift in mating distribution towards less descriminant females and the corresponding pleiotropic effect on the genome can play an important role in the population's ability to respond to changing environmental conditions. Genotypes that are better adapted to the stress conditions (e.g., those genes that confer resistance to certain diseases), are selected rapidly until the population is able to recover, especially if these genes are linked or correlated in some way to the genotypes of less discriminant females.

3.9 Population Bottlenecks and the Genetic Variability Paradox

Classical population genetic theory states that when a population is reduced in size for several generations, genetic variation is decreased due to drift

(Nei et al. 1975). For example, Mayr's (1963) model of speciation via the founder principle emphasizes the loss of genetic variability because of drift during reduced population size following the founding of a new colony. Nei et al. (1975) concluded that "when a population goes through a small bottleneck, the genetic variability of the population is expected to decline rapidly but, as soon as the population size becomes large, it starts to increase owing to new mutations". In other words, rare alleles are lost and the result is a decrease in the amount of genetic variation. However, there is much evidence now for large amounts of genetic variability even when populations are subjected to severe bottlenecks. Fitch and Atchley (1985) reported that old inbred laboratory strains of mice still carry large amounts of genetic variability. Carson (1987) reported that a highly inbred stock of *Drosophila silvestris* is chromosomally polymorphic even after nearly 10 years in the laboratory.

The differential sexual selection model discussed above provides a possible explanation of how genetic variability may not only be maintained but may actually be enhanced through the generation of novel genetic recombinants and selection for those genotypes that are better adapted for changing environmental conditions. I suggest that the sexual selection is a dynamic process influenced by density-dependent factors that enables populations to overcome harsh environmental conditions and drastic reduction in population size. Shifts in the mating distribution in response to different environmental conditions can generate a genetic environment where novel recombinants may be generated depending on the stress conditions imposed on the population. These conditions permit the population to persist until conditions improve.

3.10 The Role of Natural Hybridization

Natural hybridization between animal species, especially among insect populations, is a more widespread phenomenon than reported in the earlier literature. With the development of molecular tools (e.g., for looking at the maternally inherited mitochondrial DNA) to assay single individuals for genetic introgression (e.g., using the PCR techniques), it has been possible to better document cases of natural hybridization as well as determine the direction of introgression. It is suggested that the dynamics of the sexual selection process actually permits natural hybridization between closely related species under certain conditions. Take, for example, two species, A and B, which under normal conditions are sexually isolated so that hybridization is an extremely rare occurrence. However, when one of the species, say species B, is reduced in population size, once again there can be strong selection for less discriminant females, and therefore an increase in frequency of less discriminant females in species B. Under these conditions, females from species B may accept males from species A.

The dynamics of the sexual selection process provides the opportunity for occasional natural hybridization and permits the "leakage" of small amounts of genetic material from a related species without destroying the integrity of the separate gene pools (Kaneshiro 1990). When the population is small and there is a shift in the mating distribution towards an increase in frequency of less discriminant females, they may occasionally accept males of a related species which may be less susceptible to the stress conditions. Genes that confer such resistance to the conditions responsible for reducing the population may be strongly selected especially if such genes are linked to the genotypes of less discriminant females. Therefore, natural hybridization may be playing an important role not only in maintaining but also in replenishing genetic variability that may have been lost due to drift during small population size.

3.11 The Role of Sexual Selection in Conservation Biology

Clearly, small population sizes and bottlenecks are inherent in discussions about rare and endangered species. The dynamics of the sexual selection system as described above permit the generation of new genetic recombinants, some of which may be better adapted to the stress conditions that caused the population decline. As the population size return to normal levels, natural selection can act upon a novel genetic system.

The inference here is that if the habitat is intact, rare and endangered populations have the potential to overcome stress conditions that reduce the population to extremely low levels. The dynamics of the sexual selection system may allow selection for genotypes and phenotypes that are better suited for changes that may occur naturally in the habitat. After all, in most cases of founder event speciation, an incipient population is established by the progeny of a single fertilized female, certainly the most extreme form of a bottleneck. The genetic changes that occur in the founder colony in adapting to a new environment are similar to the kinds of changes that might occur when an established population is forced through a bottleneck in response to an ever-changing environment.

Also, the dynamics of the sexual selection system provides the mechanism whereby populations under stress might recover some of the lost genetic variability by introgressing some genetic material from a related species without destroying the integrity of the separate gene pools. If natural hybridization is indeed a more common phenomenon than previously thought possible among "good" biological species, then when endangered or threatened species are being considered, it is also necessary to understand their interaction with other species within their distributional range.

3.12 Conclusions

Several studies report empirical evidence that changes in the mating behavior of *Drosophila* species (including species other than from Hawaii) as observed in mate preference studies are related to bottleneck effects during population crashes and founder events (Ahearn 1980; Arita and Kaneshiro 1979; Dodd and Powell 1985; Ohta 1978; Powell 1978). These have been discussed and reviewed by Kaneshiro (1983, 1987) and also by Giddings and Templeton (1983) and will not be discussed in detail here. Studies of mating behavior and sexual selection in other groups of insects also support the notion that the dynamics of the sexual selection system plays an important role in the evolutionary history of the species (Löfstedt et al. 1986; Phelan and Baker 1987; Kaneshiro 1993). The results of similar studies on vertebrate systems such as on stickleback fish (McPhail 1969), mole rats (Heth and Nevo 1981), chaffinches in the Chatham Islands off the coast of New Zealand (Baker and Jenkins 1987), and pocket gophers (Bradley et al. 1991) indicate a potential, wide application of the model presented in the Chapter.

Reproductive strategies and, in particular, the complex mating system of rare and endangered species must be included as part of conservation research if we are to better understand what might be happening with the genetic material which these populations may need to recover from the conditions that made them rare and endangered in the first place. Furthermore, just because a species may be doing very well, we cannot ignore its significance in the natural community because it may be an important component of another species' genetic reservoir, especially when some stress condition forces the latter through a bottleneck situation.

Certainly, when one considers the role of the sexual selection process on the genetic consequence of a population faced with the possibility of extinction from a natural community in which it may be a keystone species, we can begin to understand the significance of the dynamics of the sexual selection system in ecosystem function. Not only does sexual selection influence the levels of genetic variability generated via novel genetic recombinants resulting from a reorganization of the genome, but also as a result of the natural hybridization that is permitted by the sexual selection model discussed above. The ability of populations to survive and adapt to changing environments, especially those that result in extreme reduction in population size, must be an important consideration of ecosystem function. It is suggested that an understanding of the sexual selection system of populations is crucial for understanding its role in the ecosystem.

Thus, the biology of small populations and the dynamics of the sexual selection process are important aspects of evolutionary biology as well as of conservation biology. A better understanding of small populations and the role of sexual selection would provide important insights into mechanisms of species formation as well as enabling us to develop more effective management plans for the rare and endangered species that can be found on island ecosystems.

Acknowledgments. I wish to thank Peter Vitousek, Hall Cushman, and the other organizers of the Workshop on Biological Diversity and Ecosystem Function on Islands held in Nassau, Bahama Islands, October 8–11, 1993, for inviting me as a participant. Thanks are also due to the Bahama National Trust for their support of the meetings. I want to thank Peter Vitousek and Henning Adsersen for reading earlier drafts of this Chapter and for their suggestions.

References

Ahearn JN (1980) Evolution of behavioral reproductive isolation in a laboratory stock of *Drosophila silvestris*. Experientia 36: 63–64
Arita LH, Kaneshiro KY (1979) Ethological isolation between two stocks of *Drosophila adiastola* Hardy. Proc Hawaii Entomol Soc 13: 31–34
Baker AJ, Jenkins PF (1987) Founder effect and cultural evolution of songs in an isolated population of chaffinches, *Fringilla coelebs*, in the Chatham Islands. Anim Behav 35: 1793–1803
Bradley RD, Davis SK, Baker RJ (1991) Genetic control of premating-isolation behavior: Kaneshiro's hypothesis and asymmetrical sexual selection in pocket gophers. J Hered 82: 192–196
Carr GD, Robichaux RH, Witter MS, Kyos DW (1989) Adaptive radiation of the Hawaiian Silversword Alliance (Compositae-Madiinae); a comparison with Hawaiian picture-winged *Drosophila*. In: Giddings LV, Kaneshiro KY, Anderson WW (eds) Genetics, speciation and the founder principle. Oxford University Press, New York, pp 79–97
Carson HL (1971) Speciation and the founder principle. Stadler Genet Symp 3: 51–70
Carson HL (1986) Sexual selection and speciation. In: Karlin S, Nevo E (eds) Evolutionary processes and theory. Academic Press, London, pp 391–409
Carson HL (1987) High fitness heterokaryotypic individuals segregating naturally within a long-standing laboratory population of *Drosophila silvestris*. Genetics 116: 415– 422
Carson HL, Hardy DE, Spieth HT, Stone WS (1970) The evolutionary biology of the Hawaiian Drosophilidae. In: Hecht MK, Steere WC (eds) Essays in evolution and genetics in honor of Theodosius Dobzhansky. Appleton-Century Crofts, New York, pp 437–543
Dalrymple GB, Silver EA, Jackson ED (1973) Origin of the Hawaiian Islands. Am Sci 61: 294–308
Darwin C (1859) The origin of species. Modern Library, New York
Darwin C (1871) The descent of man and selection in relation to sex. Modern Library, New York
Dodd DMB, Powell JR (1985) Founder-flush speciation: an update of experimental results with *Drosophila*. Evolution 39: 1388–1392
Fisher RA (1930) The genetical theory of natural selection. Clarendon Press. Oxford
Fitch RA, Atchley WR (1985) Evolution in inbred strains of mice appears rapid. Science 228: 1169–1175
Ganders FR (1989) Adaptive radiation in Hawaiian *Bidens*. In: Giddings LV, Kaneshiro KY, Anderson WW (eds) Genetics, speciation and the founder principle. Oxford University Press, New York, pp 99–112
Giddings LV, Templeton AR (1983) Behavioral phylogenies and the direction of evolution. Science 220: 372–377
Hardy DE, Kaneshiro KY (1981) Drosophilidae of Pacific Oceania. In: Ashburner M, Carson HL, Thompson JN (eds) Genetics and biology of *Drosophila*, vol 3a. Academic Press, London, pp 309–348
Heth G, Nevo E (1981) Origin and evolution of ethological isolation is subterranean mole rats. Evolution 35: 259–274
Howarth FG, Mull WP (1992) Hawaiian insects and their kin. University Hawaii Press, Honolulu
Kaneshiro KY (1980) Sexual isolation, speciation, and the direction of evolution. Evolution 34: 437–444
Kaneshiro KY (1983) Sexual selection and direction of evolution in the biosystematics of Hawaiian Drosophilidae. Annu Rev Entomol 28: 161–178
Kaneshiro KY (1987) The dynamics of sexual selection and its pleiotropic effects. Behav Genet 17: 559–569
Kaneshiro KY (1988) Modes of speciation in Hawaiian *Drosophila*. BioScience 38: 258–263
Kaneshiro KY (1989) The dynamics of sexual selection and founder effects in species formation. In: Giddings LV, Kaneshiro KY, Anderson WW (eds) Genetics, speciation and the founder principle. Oxford University Press, New York, pp 279–296

Kaneshiro KY (1990) Natural hybridization in *Drosophila*, with special reference to species from Hawaii. Can J Zool 68: 1800–1805
Kaneshiro KY (1993) Introduction, colonization, and establishment of exotic insect populations: fruit flies in Hawaii and California. Am Entomol 39: 23–29
Kaneshiro KY, Boake CRB (1987) Sexual selection and speciation: issues raised by Hawaiian *Drosophila*. Trends Ecol Evol 2: 207–212
Kaneshiro KY, Giddings LV, (1987) The significance of asymmetrical sexual isolation and the formation of new species. Evol Biol 21: 29–43
Kirkpatrick M (1982) Sexual selection and the evolution of female choice. Evolution 36: 1–12
Lande R (1981) Models of speciation by selection on polygenic traits. Proc Natl Acad Sci 78: 3721–3725
Lande R (1982) Rapid origin of sexual isolation and character divergence in a cline. Evolution 36: 213–233
Löfstedt C, Herrebout WM, Du JW (1986) Evolution of the ermine moth pheromone tetradecyl acetate. Nature 323: 621–623
Mayr E (1942) Systematics and the origin of species. Columbia University Press, New York
Mayr E (1963) Animal species and evolution. Harvard University Press, Cambridge, MA
Mayr E (1972) Sexual selection and natural selection. In: Campbell B (ed) Sexual selection and the descent of man, 1871–1971, Aldine, Chicago, pp 87–104
Mcphail JD (1969) Predation and the evolution of a stickleback (*Gasterosteus*). J Fish Res Board Can 26: 3183–3208
Nei M, Maruyama T, Chakraborty R (1975) The bottleneck effect and genetic variability in populations. Evolution 29: 1–10
O'Donald P (1977) Theoretical aspects of sexual selection. Theor Popul Biol 12: 298–334
O'Donald P (1980) Genetic models of sexual selection. Cambridge University Press, Cambridge
Ohta AT (1978) Ethological isolation and phylogeny in the grimshawi species complex of Hawaiian *Drosophila*. Evolution 32: 485–492
Olson SL, James HF (1982) Fossil birds from the Hawaiian Islands: evidence for wholesale extinction by man before western contact. Science 217 (4560): 633–635
Phelan LP, Baker TC (1987) Evolution of male pheromones in moths: reproductive isolation through sexual selection? Science 235: 205–207
Powell JR (1978) The founder-flush speciation theory: an experimental approach. Evolution 32: 465–474
Throckmorton LH (1975) The phylogeny, ecology and geography of *Drosophila*. In: King RC (ed) Handbook of genetics. Plenum New York, pp 431–469
West-Eberhard MJ (1983) Sexual selection, social competition, and speciation. Q Rev Biol 58: 155–183
Wilcox BA (1980) Insular ecology and conservation. In: Soulé ME, Wilcox BA (eds) Conservation biology: an evolutionary-ecological perspective. Sinauer, Sunderland, MA, pp 95–117

4 Patterns of Diversity in Island Plants

U. ELIASSON

4.1 Introduction

Islands form smaller and less complicated ecosystems than do neighbouring continents, and for this reason are interesting objects in the study of evolution. These small ecosystems tend to be easily disturbed and vulnerable, however, the smaller and more isolated they are. The geological history may be of great importance for the composition of the biological system and the evolutionary direction on an island. Many islands were once connected to a continent, whereas others were created through volcanic activity on the sea floor and have always been surrounded by water. Biologists mostly refer to the two types as continental and oceanic, respectively, but there are islands that do not fit into either of these categories due to complicated geological events in the past. Although islands in the normal sense are pieces of land surrounded by water, restricted habitats surrounded by different environmental conditions may be regarded as islands in a strict biological sense.

The literature on island biology is rather extensive. This chapter is an attempt to summarize briefly some general features of the morphological diversity exhibited by plants on isolated oceanic islands. Treated in greater detail are some notable examples of evolution on the Hawaiian Islands and the Galápagos archipelago.

4.2 Disharmony and Endemism

Isolated oceanic islands are mostly characterized by disharmonic floras with a varying portion of their species being endemic. Disharmony means that the composition and proportion of different families differ more or less strongly from the flora of the source region, mostly the nearest continent. Floras as well as faunas on isolated islands are generally the result of long-distance dispersal and it is natural that plants on such islands belong to families characterized by good dispersal ability, even if the species now present on the island may have lost that ability. Plant families important on a continent may be poorly represented or lacking on an outlying island, simply because representatives of the family never managed to get there and become established. Examples are the monocotyledonous

Department of Systematic Botany, Göteborg University, Carl Skottsbergs Gata 22, S-413 19 Göteborg, Sweden

Ecological Studies, Vol. 115
Vitousek et al. (eds) Islands

families Araceae, Bromeliaceae, and Cyclanthaceae, which are important in Ecuador and Peru, but are represented by only one single endemic species of Bromeliaceae on the Galápagos Islands.

The proportion of endemic species varies strongly. Endemism is dependent on several factors, the most important of which are the ecological breadth of the island and the degree of isolation. The ecological breadth in turn depends on size, age, geology, and edaphic conditions. Isolated islands are generally, but not always, poorer in families and genera than a continental area of similar size. The number of species in an island flora is generally positively correlated with the size of the island and the proximity to the source region. Relationships among various factors governing the species number on an island were discussed in the famous book by MacArthur and Wilson (1967). Several variables, however, are not easily expressed mathematically. As discussed by van Balgooy (1969), several variables, although appearing simple at first sight, are not easily defined. The surface area available to colonizers is larger on a high and mountainous island than on a low island, even if both may appear to be the same size when measured on a map. The area of an island may have been larger during earlier periods. The distance to a source region may also be different today than it was in the past, and islands now sunken may have functioned as stepping stones. Even if there are islands that for various reasons are difficult to fit into the formulas presented by MacArthur and Wilson (1967), the book is still a good starting point for analysis of almost any island flora.

4.3 Adaptive Radiation

If an island is rich in habitats, a newcomer may gradually give rise to several species through adaptive radiation, with each species occupying its own ecological niche. Mutations that would be disadvantageous in the source area may survive in a new environment where competition and perhaps predation pressure are weaker. Habitat diversification is a basic condition for adaptive radiation. Thus, endemism is a rare phenomenon on low coral islands. This is due not only to the uniform environment but also to the fact that the plants established are likely to have arrived repeatedly by sea; repeated arrival of propagules of the same species will swamp possible genetic differences in the plants already on the island. Only on higher islands may so-called beach species give rise to new taxa adapted to new habitats.

Although adaptive radiation has been encountered in many different families, it is most beautifully demonstrated by widespread families that are weedy in the sense that they have a good dispersal ability and are pioneer plants of disturbed habitats. Among such families, the Asteraceae stands out as the most successful and has given rise to a large number of endemic species as well as genera on isolated islands. Reasons for the success of this family are to be sought in a combination of special features: (1) the family is large and widespread, (2) the bulk of the family comprises weedy species, (3) the majority of species possess fruits adapted for dispersal over long distances by wind or animals, and (4) the family is plastic in terms of evolution and adaptability. The great morphological and evolutionary plasticity has resulted in insular growth forms that are very different from their presumed herbaceous ancestors. Such forms include everything from herbaceous shrubs with lignification only at base to true trees 10 m tall or more. Tall herbs with a basal leaf rosette, as well as forms with a large rosette on top of a trunk, are also found. This morphological plasticity is exhibited by several different tribes within the large family (Carlquist 1965, 1974). In the Juan Fernández Islands there are five endemic genera representing three different tribes, all of which have developed a strong degree of lignification and a more or less tree-like habit (Carlquist 1965; Sanders et al. 1987). These genera are *Centaurodendron* and *Yunquea* of the tribe Cardueae, *Dendroseris* of the Lactucae, and *Rhetinodendron* and *Robinsonia* of the Senecioneae. The eleven species of *Dendroseris* and seven species of *Robinsonia* have evolved at the tetraploid level without changes in chromosome number. Speciation is believed to have taken place by geographical isolation of peripheral populations (Sanders et al. 1987). Surprisingly, none of the endemic genera on Juan Fernández belongs to the Heliantheae, a tribe that demonstrates such conspicuous examples of adaptive radiation in Hawaii and the Galápagos Islands.

The excellent dispersibility of spores makes ferns relatively richly represented on tropical and subtropical oceanic islands. With few exceptions, like Hawaii, where about 65% of the pteridophytes are endemic, the proportion of endemic taxa is generally small. This may point to a low degree of evolutionary plasticity but, more likely, is due to the genetic system of the group. Oceanic islands are predominantly settled by isosporous species with hermaphrodite and self-compatible gametophytes (Ehrendorfer 1979). The result of self-fertilization in a haploid gametophyte leads to homozygous sporophytes whatever the degree of heterozygosity of the sporophyte that produced the spore. Heterozygosity is an important factor in the diversification of a taxon into new forms through adaptive radiation.

4.4 Morphological Features of Island Plants

4.4.1 Island Woodiness

The alleged tendency of plants of predominantly herbaceous families to become lignified and form shrubs and trees on islands has been discussed extensively in a series of publications by Carlquist (1965, 1969, 1970, 1974), who has interpreted woodiness in some cases to result from climatic conditions. In a mild and uniform environment where no drastic climatic changes influence flowering or growth processes, continued growth and flowering may be advantageous. A continued growth of stems in otherwise short-stemmed plants may result in leafy crowns being pushed upwards on top of a thick stem (Carlquist 1965) and might at least partly explain why the rosette-tree habit is relatively common on islands with a uniform climate. In Hawaii, this habit is common among members of the Asteraceae and the Campanulaceae-Lobelioideae. Woodiness as a characteristic feature of island plants is controversial, however. Van Steenis (1973) objected to the ecological explanation that a moderate climate would permit continued growth, and suggested as an explanation that empty niches on barren islands would offer colonizers the opportunity to develop woody life forms. Most families of flowering plants probably possess the ability to develop woody growth when conditions are right.

The supposition that woodiness in several predominantly herbaceous families would be a feature restricted to islands lingers from a time when the interior of continents was still insufficiently explored. Woody representatives of predominantly herbaceous groups do occur on continents as well. Thus, there are several lignified genera of Campanulaceae-Lobelioideae in western South America, and lignified species of *Senecio* (s. lat.) and *Lobelia* are found in island-like, high-elevation habitats of East Africa. Nevertheless, woody representatives are conspicuous elements in floras of many volcanic islands. The explanation does not have to be a simple one, and I believe that Carlquist's as well as van Steenis' hypotheses have their relevance on different archipelagos. It should be stressed, however, that woodiness is not a simple character but may be of different kinds. The degree of lignification may vary strongly. Some species of the rosette-bearing stem habit are not woody in the true sense of the word. Thus, a thick stem may be succulent rather than woody, such as in the genus *Aeonium* (Crassulaceae) of the Canary Islands. Several genera of the centrospermous families possess the potential of continued growth through the ability to develop new cambia, so-called anomalous secondary thickening. This has resulted in woody species of genera such as *Chenopodium* (Hawaii, Juan Fernández), *Charpentiera* (Hawaii, Austral Islands, Cook Islands), and *Alternanthera* (Galápagos Islands) (Fig. 4.1, 4.2). There are genera of quite other families that seem to have a potential for some kind of woodiness inherited. An example is the woody stem that has evolved in three species of *Plantago* on widely separated archipelagos (*P. princeps* in Hawaii, *P. fernandezianum* on Juan Fernández, and *P. canariensis* on the Canary Islands).

The rosette-tree habit of tree ferns, a plant group forming an important element in the floras of some islands, is due to the special morphology of this group of plants and is not a response to island conditions. The relatively frequent occurrence of tree ferns on isolated islands is regarded as a reflection of their excellent dispersibility.

4.4.2 Loss of Dispersibility

Good dispersal ability is a prerequisite for getting to an isolated island. Once there, however, possessing diaspores adapted to long-distance dispersal would be disadvantageous, since the majority of them would risk being dropped over the

Fig. 4.1. *Alternanthera echinocephala*, Santa Cruz, Galápagos Islands. This endemic species is a good example of woodiness among the centrospermous families. The specimen is ca 1.7 m tall. (Photo: U. E., Dec. 1980)

open sea. In fact, many island species show a strikingly reduced dispersal ability when compared with their presumed ancestors (Carlquist 1966, 1974). This loss of dispersibility may consist in reduction of attachment devices on fruits and seeds or in increased size and weight of diaspores. Possession of larger and heavier

Fig. 4.2. *Alternanthera echinocephala.* Basal part of stem with so-called anomalous secondary growth. (Photo: U. E., Dec. 1980)

seeds means that the diaspores will mostly be dropped in the vicinity of the mother plant, hence, in an environment that is likely to be favourable for the seedlings. Large seeds contain more endosperm for nourishment of the early critical stages of the new plant. Among the Asteraceae there are numerous examples of reduced dispersibility in island species. The genus *Bidens* has several widespread species, the fruits of which have barbed awns that efficiently aid in epizoic dispersal. In endemic species of the same genus in Hawaii, the pappus is strongly reduced or lacking and the shape of the fruit is more or less curved. These fruits are no longer suited for epizoic dispersal. Several species of the same family on Juan Fernández have strongly enlarged achenes with reduced bristles (Carlquist 1966), the shape of the fruit is irregular in some species, and the achenes may remain in the flower-head for long periods of time.

4.4.3 Changes in Reproductive Biology

Among other features ascribed to island floras should be mentioned changes in the reproductive biology, especially adoption of various modes to promote outcrossing. Again, this is a field that has been excellently surveyed by Carlquist (1974). Dioecism may be regarded as the ultimate mode for the promotion of outcrossing, and the percentage of dioecious species is mostly considerably higher in island floras than on continents. The proportion of dioecious species on continental areas rarely exceeds 3%, but very often exceeds 10% in island floras. The proportion of dioecious species in the indigenous flora of Hawaii has been estimated at about 27% (Carlquist 1974), a figure that may be exaggerated.

4.5 Selected Examples of Adaptive Radiation in Hawaii and the Galápagos Islands

Hawaii is the prime illustration of an oceanic archipelago with magnificent examples of adaptive radiation and other insular flora characteristics (Carlquist 1970). It is important to note, however, that only a small portion of the presumed original colonists have diversified and given rise to a substantial number of species through adaptive radiation. Such evolutionarily successful genera are found in several systematically scattered families, for example the Gesneriaceae, Lamiaceae, Rubiaceae, Rutaceae, Thymelaeaceae, but the most outstanding examples are found in the Asteraceae and the Campanulaceae-Lobelioideae. There are six endemic genera of Asteraceae and many endemic species of nonendemic genera of the same family. Out of ca. 91 species of Asteraceae native to Hawaii all but one (*Adenostemma lavenia*) are endemic (Wagner et al. 1990). The widespread genus *Bidens* (Heliantheae-Coreopsidinae) has ca. 19 endemic species in Hawaii, besides a few naturalized species, and possesses several features characteristic of insular plants (Ganders and Nagata 1990). Species of the endemic Asteraceae genera are all examples of adaptive radiation, made possible through a wide diversity of habitats. These genera are *Hesperomannia* (3 spp.) of the Mutisieae, *Remya* (3 spp.) of the Astereae, *Argyroxiphium* (5 spp.), *Dubautia* (21 spp.) and *Wilkesia* (2 spp.) of the Heliantheae-Madiinae, and *Lipochaeta* of the Heliantheae-Ecliptinae. The four latter genera will be treated in somewhat greater detail.

In the Galápagos Islands, which are geographically isolated, although far less so than Hawaii, there are six endemic genera of vascular plants, two of which are cacti and four are genera of the Asteraceae. Three of the latter genera belong to the tribe Heliantheae. Endemic genera are *Darwiniothamnus* (3 spp.) of the Astereae, *Lecocarpus* (3 spp.) of the Heliantheae-Melampodiinae, *Macraea* (1 sp.) of the Heliantheae-Ecliptinae, and *Scalesia* (15 spp.) of the Heliantheae-Helianthinae). The three latter genera will be further commented on below.

4.5.1 The Hawaiian Silversword Alliance

One of the most outstanding examples of adaptive radiation in island plants is found in the so-called Hawaiian silversword alliance. These are plants of the Asteraceae-Heliantheae-Madiinae, comprising altogether 28 species in three genera, *Argyroxiphium*, *Dubautia*, and *Wilkesia*, all endemic to the Hawaiian Islands (Carr 1990). An extensive monograph of all three genera has been published by Carr (1985). In a subsequent series of papers Carr and coworkers have demonstrated the morphological, cytological, ecological, and physiological diversity of the group.

The morphological diversity is extremely wide in response to different habitats. The different life-forms include mat-forming subshrubs, cushion plants, rosette shrubs, trees, and vines. Species of *Argyroxiphium* and *Wilkesia* are rosette shrubs, the rosettes being sessile in some species or placed on top of woody stems up to 5 m tall in other species. Most species of *Dubautia* are shrubs of varying habit. Two species of *Dubautia* form true trees up to 5–8 m tall with trunks up to 0.5 m in diameter, and one species is a liana climbing up to 8 m into the canopy of large trees (Robichaux et al. 1990). Leaf length in the entire alliance varies from 5 mm up to 0.5 m. Species of *Argyroxiphium* and *Wilkesia* have narrowly ligulate or linear leaves, whereas linear as well as broader leaves of various shapes occur in *Dubautia* (Carr 1985). Physiological diversity is often coupled with differences in leaf anatomy and is illustrated by differences in leaf turgor maintenance capacity. This enables species to inhabit a wide range of ecological niches, from dry and hot lava fields near sea level to wet forests and bogs at altitudes up to 3700 m (Carr and Kyhos 1986). Differences in local distributions among sympatrically occurring species may reflect differences in anatomical and physiological features (Robichaux et al. 1990).

Despite the strikingly different habit and morphology, the three genera in the silversword alliance have proved to be remarkably coherent genetically (Carr and Kyhos 1986). There is a high frequency of spontaneous hybridization. It is easy to produce artificial hybrids and the hybrids are capable of producing backcross progeny. Many of the genomes that have been distinguished have been found to be common to a large number of species (Carr and Kyhos 1986). Evidence from studies using cytogenetic, electrophoretic, and molecular approaches indicates that the origin of the whole group could be traced to a single colonizing ancestor (Carr et al. 1989; Baldwin et al. 1990; Robichaux et al. 1990).

Anatomical and morphological evidence indicates taxonomic alignment of the silversword alliance with species of the Madiinae of western North America (Carlquist 1959). The North American linkage has been strongly reinforced by data from chloroplast DNA restriction-site comparisons (Baldwin et al. 1991). The data demonstrate that the chloroplast DNA and at least one nuclear genome of the tetraploid silversword alliance arose from an extant Californian lineage that includes diploid, montane, perennial herbs of the genera *Raillardiopsis* and *Madia*. The close genetic affinity has been illustrated by successful attempts to cross artificially a species of the Hawaiian genus *Dubautia* with species of the Californian genera. All progeny obtained were vigorous hybrids. Attempted crosses between species of the silversword alliance on the one hand and North American Madiinae of genera other than the two mentioned have failed (Baldwin et al. 1991). Few intergeneric crosses have proven possible among mainland Madiinae.

The North American linkage of the Hawaiian silversword alliance is phytogeographically noteworthy in view of the preponderantly Pacific and East Asian affinity of the Hawaiian flora. Sticky bracts and appendages on the fruit may have made possible the long-distance dispersal of a North American diaspore, perhaps attached to the plumage of a migratory bird.

4.5.2 *Lipochaeta* in Hawaii

Lipochaeta is a genus of the Asteraceae-Heliantheae-Ecliptinae (Stuessy 1977) endemic to the Hawaiian Islands. Gardner (1979) recognized 27 species, but the number has more recently been reduced to 20 (Wagner et al. 1990). All species but one are shrubby perennials. There is convincing evidence that the genus is biphyletic. It can be divided into two certainly natural and well-delimited sections (Gardner 1976, 1979), characterized by morphological as well as chromosomal and chemical features. The nominate section comprises six species; all species are tetraploids with 26 chromosome pairs, the disk florets have mostly 4-lobed corollas, and the species can synthesize flavones as well as flavonols. The section *Aphanopappus* includes 14 species; all species are diploids with 15 chromosome pairs, the disk florets have mostly 5-lobed corollas, and the species can synthesize flavonols but not flavons. Members of the same section are highly interfertile, whereas the genetic barrier between the two sections is very strong (Rabakonandrianina 1980). The genetic barrier is primarily due to the difference in ploidy level.

The close relationship between *Lipochaeta* and *Wedelia* has been demonstrated in a study involving intergeneric hybridization (Rabakonandrianina and Carr 1981). The section *Lipochaeta* is believed to have had an allopolyploid origin with *Wedelia biflora* DC. or a closely related species, and an unknown species with 11 chromosome pairs (Rabakonandrianina 1980; Wagner et al. 1990). The section *Aphanopappus* has also been demonstrated to be closely related to *Wedelia biflora* (Wagner et al. 1990). There is convincing evidence that each of the two sections of *Lipochaeta* originated from ancestors of *Wedelia* sensu lato that independently colonized Hawaii. As to dispersal mechanisms, it has been suggested (Gardner 1976) that the thick corky outer wall of the achenes of *Wedelia biflora* may aid in flotation and that this or a closely related species may have reached Hawaii from somewhere in the Pacific. Similar corky achenes, although with thinner walls, occur in a few species of *Lipochaeta* (Gardner 1976, 1979).

It is a frequently repeated phenomenon on oceanic islands that small isolated populations initiate a rapid evolution where a wide diversity of habitats is available. This has been demonstrated in *Bidens* in the Hawaiian Islands where taxa exhibit a great morphological and ecological diversity (Gillett and Lim 1970). The 19 endemic species are believed to have originated from a single ancestor and are a beautiful example of adaptive radiation (Ganders and Nagata 1984). Although there are no genetic barriers, natural hybrids are rarely formed since species are adapted to different habitats and only rarely meet. Similar conditions appear to hold for *Lipochaeta* where geographical and/or ecological conditions keep the species apart. The diploid species referred to the section *Aphanopappus* are mostly geographically restricted to one island and often to a particular valley or mountain range (Gardner 1976, 1979). Only a few species have wider distributions. Species of the section *Lipochaeta* are more widely distributed.

4.5.3 Lobelioideae in Hawaii

Although the Asteraceae are outstanding in offering striking examples of adaptive radiation on oceanic islands, the same phenomenon does occur in other families as well. Examples are the genera *Coprosma* (Rubiaceae), *Cyrtandra* (Gesneriaceae), and *Melicope* (= *Pelea*; Rutaceae) in Hawaii with ca. 13, 53, and 47 species, respectively (Wagner et al. 1990). One of the most magnificent examples of adaptive radiation in a family outside the Asteraceae is demonstrated by the subfamily Lobelioideae of the Campanulaceae in Hawaii. Besides several endemic species of *Lobelia*, six endemic genera have been recognized, namely (number of species in paranthesis), *Clermontia* (21), *Cyanea* (52), *Delissea* (9), *Rollandia* (8), *Brighamia* (2), and *Trematolobelia* (4) (Lammers 1990b, 1991). These endemics exhibit an extreme morphological diversity in growth form, leaf shape, and shape and colour of corolla. The first four genera are believed to be derived from a single common ancestor and share several features which are unique among the Lobelioideae (Lammers 1990b). Their close affinities have recently been illustrated by the formal incorporation of *Rollandia* within *Cyanea* (Lammers et al. 1993). Wide diversity in corolla shape is believed to be an adaptation to the beak shapes of specific species of endemic nectar-feeding birds that have been important in the pollination. Roughly 25% of the endemic lobelioid species have become extinct during the past century, and the percentage is even higher for the endemic birds. Although self-pollination is possible in the

Lobelioideae and several plant species may be confined to this mechanism when the pollinators are gone, exclusive self-pollination will presumably in the long run decrease genetic diversity and threaten the existence of the species (Carlquist 1965). The importance of insect pollination in the endemic lobelioids is not quite clear. It is tempting to assume, however, that the splendid diversity in corolla shapes results from adaptation and coevolution with the endemic drapanidid birds.

4.5.4 *Scalesia* in the Galápagos

The genus *Scalesia* (Compositae-Heliantheae) is endemic to the Galápagos Islands. It is one of the best examples of adaptive radiation in the archipelago and comprises about 15 species (Eliasson 1974; Hamann and Wium-Andersen 1986) and altogether 21 taxa. The genus is widespread in the archipelago and is absent from only three of the major islands (Española, Genovesa, Marchena) and from the remote islet of Darwin. Most species have restricted distributions and occur on one single island. All species are woody. The majority of species form shrubs 0.3–2 m tall and several grow on relatively fresh lava fields. Three species form trees, two (*S. cordata* Stewart, *S. pedunculata* Hook.f.) reaching 10 m or more in height. There is a striking diversity in the appearance of leaves, from entire blades of varying shape and with different types and amount of vestiture to lobed and finely dissected. Flower-heads are homogamous in most species with bisexual disk-flowers. One species (*S. affinis* Hook. f.) has radiate heads with bisexual disk-flowers and sterile ray-flowers. In some species with predominantly homogamous heads there is an incipient differentiation of peripheral disk-flowers into rays. These corollas become irregularly cleft and oblique with a concomitant reduction of stamens and gynoecium.

The genus has been referred to the subtribe Helianthinae on morphological and cytological evidence (Eliasson 1974; Robinson 1981). Eliasson (1974) regarded *Helianthus* and *Viguiera* as the mainland genera closest allied to *Scalesia* but refrained from tracing a more detailed origin due to the doubtful taxonomic delimitation of the two former genera. Common characters emphasized were the shape and venation of leaves, presence of sterile ray-florets, laterally compressed achenes, and paleae that are folded and embracing the achenes. Trifid paleae are characteristic of *Scalesia* but a tendency towards three-cleft paleae may be found in at least some species of *Viguiera*. There have been profound taxonomic changes in generic delimitations since the publication of Eliasson's monograph. In Watson's (1929) monograph of *Helianthus*, the genus had a disjunct distribution with the largest number of species in North America and a smaller number in the Andean region of South America. Heiser (1957) regarded the genus as diphyletic, suggesting that the species had arisen from *Viguiera* independently in North and South America. The affinity of the South American species with *Viguiera* rather than *Helianthus* was further emphasized by Robinson (1979), who referred the former South American species of *Helianthus* to the new genus *Helianthopsis*. More recently, Panero (1992) has regarded this taxon as congeneric with *Pappobolus*. This genus extends from southern Colombia to central Peru, with 30 of the 38 species being endemic to Peru. The greatest concentration of species is found in the numerous valleys of the Marañón River Basin just south of the Huancabamba River gorge (Panero 1992).

A recent study (Schilling et al. 1994) based on chloroplast DNA restriction site analysis indicates that *Scalesia* is a sister group of the Andean genus *Pappobolus*. This view may also receive support from other cytological and morphological features. The basic chromosome number of $x = 17$ is the same for the two genera, but the mainland genus is diploid whereas all species of the Galápagos genus are tetraploid with $2n = 68$. Sterility barriers appear to be weak, and it is possible that natural hybridization has been an important factor in the diversification.

Scalesia on the Galápagos Islands is another example of reduced dispersibility. Most species lack appendages on the fruits or have only one or rarely two short straight awns. A moderately well-developed pappus is present in only one species, the arboreous *S. cordata*, but the appendages are poorly suited as devices for efficient dispersal. All species probably drop their achenes close to the mother tree. Although many species of *Pappobolus*, the mainland genus suggested to be the sister group to *Scalesia* (Schilling et al. 1994), have only two awns, which may be caducous, other species may have a more persistent pappus with awns as well as squamellae (Panero 1992).

4.5.5 *Macraea* in the Galápagos

The monotypic genus *Macraea* is endemic to the Galápagos Islands where it is widespread and forms shrubs 1–2 m tall. It is a genus of the Compositae-Heliantheae. It was apparently overlooked by Stuessy (1977) in his revised system of the Heliantheae but would key to the subtribe Ecliptinae. Although originally described as a distinct genus (Hooker 1847), it was later transferred (Gray 1862) to the otherwise strictly Hawaiian genus *Lipochaeta*. A thorough morphological study (Harling 1962) demonstrated important differences between the Galápagos species and Hawaiian material of *Lipochaeta*. A further difference was found in the chromosome number, that was found to be $2n = 28$ in the Galápagos species (Eliasson 1984) whereas in *Lipochaeta* species are either diploids with the gametic number $n = 15$ or tetraploids with $n = 26$. Harling (1962) concluded that *Macraea* was more closely allied to *Wedelia* and *Aspilia* than to the Hawaiian genus and emphasized the structure of the pappus and the basal construction of the achenes into a sterile stipe as the most important similarities between *Macraea* and several species of *Wedelia*, among them *W. trilobata* (L.) Hitchc. Recent molecular studies corroborate the results based on morphological investigations that the closest affinities of *Macraea* are found in *Wedelia*. These studies have shown *Macraea* to be the sister taxon of *W. trilobata* (Panero, pers. comm.). This is well in line with the findings that insular taxa are often derived from widespread continental genera.

Although *Macraea* is regarded as a monotypic genus there is an incipient differentiation resulting in morphologically distinct populations on different islands.

Differences are found in general habit, leaf size, and structure of flower heads. On the island of Floreana, the flower heads are commonly radiate with pistillate ray-flowers with ligulate corollas. On most islands, the heads are predominantly homogamous with all florets being bisexual with actinomorphic corollas. On several islands (Floreana, Isabela, Santiago) populations are found where the outermost disk-flowers are in different stages of metamorphosis towards ligulate ray-corollas (Harling 1962). In contrast to the ligulate ray-corollas of, e.g., *Scalesia*, those of *Macraea* seem to have retained their sexual function and ability to produce viable achenes.

4.5.6 *Lecocarpus* in the Galápagos

The genus *Lecocarpus* belongs in the Heliantheae-Melampodiinae and comprises three shrubby species restricted to one island each in the southeastern part of the Galápagos archipelago (Adsersen 1980). The best known species, *L. pinnatifidus* Decaisne, occurs in scattered localities on the island of Floreana (Eliasson 1971; Adsersen 1980). Leaves of mature specimens are deeply lobed with lobes entire or again lobed to deeply pinnatifid. *Lecocarpus lecocarpoides* (Robins. & Greenm). Cronquist & Stuessy is a species of Española and the adjacent islet of Gardner. The basic pattern of leaf lobation resembles that in the previous species. *Lecocarpus darwinii* Adsersen is restricted to San Cristóbal, the easternmost island in the archipelago. The leaves are serrate and resemble those of seedlings of *L. pinnatifidus*.

Fig. 4.3. *Lecocarpus pinnatifidus*, Floreana, Galápagos Islands. The species illustrates a change in dispersal ability. Only ray-flowers are fertile. A bract tightly encloses the fruit and forms a peltate wing. (Photo: U. E., Jan. 1981)

Fig. 4.4. *Lecocarpus pinnatifidus*. Diaspores on the ground in immediate vicinity of the mother plant. (Photo: U. E., Jan. 1981)

This may be a case of so-called pedomorphosis, the retention of juvenile characters, but further studies are needed. Pedomorphosis is a phenomenon sometimes encountered in island biology, although in no way restricted to islands, and has been demonstrated in the endemic Campanulaceae-Lobelioideae of the Hawaiian Islands (Lammers 1990a).

Lecocarpus illustrates a change in dispersal ability. Fruits of this and allied genera are tightly enclosed by an indurated bract, the whole structure serving as a dispersal unit. *Lecocarpus darwinii* and *L. lecocarpoides* possess prickly diaspores (Adsersen 1980), a feature shared with *Acanthospermum*, a closely allied genus on the mainland represented by one nonendemic but possibly indigenous species in Galápagos. *Lecocarpus pinnatifidus* has unarmed diaspores provided with a peltate wing (Fig. 4.3). This reshaped diaspore apparently demonstrates a change in dispersal strategy, it has been suggested as a response to the lack of fur-bearing animals. The wing will make the fruit fall more slowly to the ground and, hence, increase the chance of off-drift from the immediate vicinity of the mother plant (Fig. 4.4). Relatively few achenes are produced since only the ray-flowers are female-fertile.

Acknowledgment. I thank Dr. Jose L. Panero for letting me use some so far unpublished information regarding the affinities of *Macraea.*

References

Adsersen H (1980) Revision of the Galápagos endemic genus *Lecocarpus* (Asteraceae). Bot Tidsskr 75: 63–76

Baldwin BG, Kyhos DW, Dvorák J (1990) Chloroplast DNA evolution and adaptive radiation in the Hawaiian silversword alliance (Asteraceae-Madiinae). Ann Mo Bot Gard 77: 96–109

Baldwin BG, Kyhos DW, Dvorák J, Carr GD (1991) Chloroplast DNA evidence for a North American origin of the Hawaiian silversword alliance (Asteraceae). Proc Natl Acad Sci USA 88: 1840–1843
Carlquist S (1959) Studies in Madiinae: anatomy, cytology, and evolutionary relationships. Aliso 4: 171–236
Carlquist S (1965) Island life. The Natural History Press, New York, 451 pp
Carlquist S (1966) The biota of long-distance dispersal. II. Loss of dispersibility in Pacific Compositae. Evolution 20: 30–48
Carlquist S (1969) Wood anatomy of Goodeniaceae and the problem of insular woodiness. Ann Mo Bot Gard 56: 358–390
Carlquist S (1970) Hawaii, a natural history. The Natural History Press, New York, 463 pp
Carlquist S (1974) Island biology. Columbia University Press, New York, 660 pp
Carr GD (1985) Monograph of the Hawaiian Madiinae (Asteraceae): *Argyroxiphium*, *Dubautia*, and *Wilkesia*. Allertonia 4: 1–123
Carr GD (1990) *Argyroxiphium*, *Dubautia*, *Wilkesia*. In: Wagner WL, Herbst DR, Sohmer SH (eds) Manual of the flowering plants of Hawaii, Vol 1. University of Hawaii Press, Honolulu
Carr GD, Kyhos DW (1986) Adaptive radiation in the Hawaiian silversword alliance (Compositae-Madiinae). II. Cytogenetics of artificial and natural hybrids. Evolution 40: 959–976
Carr GD, Robichaux RH, Witter MS, Kyhos DW (1989) Adaptive radiation of the Hawaiian silversword alliance (Compositae-Madiinae): a comparison with Hawaiian picture-winged *Drosophila*. In: Giddings LV, Kaneshiro KY, Anderson WW (eds) Genetics, speciation, and the founder principle. Oxford University Press, New York, pp 79–97
Ehrendorfer F (1979) Reproductive biology in island plants. In: Bramwell D (ed) Plants and islands. Academic Press, London, pp 293–306
Eliasson U (1971) Studies in Galápagos plants. X. The genus *Lecocarpus* Decaisne. Sven Bot Tidskr 65: 245–277
Eliasson U (1974) Studies in Galápagos plants. XIV. The genus *Scalesia* Arn. Opera Bot 36: 1–117
Eliasson U (1984) Chromosome number of *Macraea laricifolia* Hooker fil. (Compositae) and its bearing on the taxonomic affinity of the genus. Bot J Linn Soc (Lond) 88: 253–256
Ganders FR, Nagata KM (1984) The role of hybridization in the evolution of *Bidens* on the Hawaiian Islands. In: Grant WF (ed) Biosystematics. Academic Press Canada, Ontario, pp 179–194
Ganders FR, Nagata KM (1990) Bidens. In: Wagner WL, Herbst DR, Sohmer SH (eds) Manual of the flowering plants of Hawaii. University of Hawaii Press, Honolulu, pp 267–283
Gardner RC (1976) Evolution and adaptive radiation in *Lipochaeta* (Compositae) of the Hawaiian Islands. Syst Bot 1:383–391
Gardner RC (1979) Revision of *Lipochaeta* (Compositae: Heliantheae) of the Hawaiian Islands. Rhodora 81: 291–343
Gillett GW, Lim EKS (1970) An experimental study of *Bidens* (Asteraceae) in the Hawaiian Islands. Univ Calif Publ Bot 56: 1–63
Gray A (1862) Characters of some Compositae in the United States South Pacific exploring expedition under Captain Wilkes, with observations by Asa Gray. Proc Am Acad Arts Sci 5: 114–146
Hamann O, Wium-Andersen S (1986) *Scalesia gordilloi* sp. nov. (Asteraceae) from the Galápagos Islands, Ecuador. Nord J Bot 6: 35–38
Harling G (1962) On some Compositae endemic to the Galápagos Islands. Acta Horti Bergiani 20: 63–120
Heiser CB (1957) A revision of the South American species of *Helianthus*. Brittonia 8: 283–295
Hooker JD (1847) An enumeration of the plants of the Galápagos archipelago with descriptions of those which are new. Trans Linn Soc Lond 20: 163–233
Lammers TG (1990a) Sequential paedomorphosis among the endemic Hawaiian Lobelioideae (Campanulaceae). Taxon 39: 206–211
Lammers TG (1990b) Campanulaceae. In: Wagner WL, Herbst DR, Sohmer SH (eds) Manual of the flowering plants of Hawaii. University of Hawaii Press, Honolulu, pp 420–489
Lammers TG (1991) Systematics of *Clermontia* (Campanulaceae-Lobelioideae). Syst Bot Monogr 32: 1–97

Lammers TG, Givnish TJ, Sytsma KJ (1993) Merger of the endemic Hawaiian genera *Cyanea* and *Rollandia* (Campanulaceae: Lobelioideae). Novon 3: 437–441
MacArthur RH, Wilson EO (1967) The theory of island biogeography. Princeton University Press, Princeton, 203 pp
Panero JL (1992) Systematics of *Pappobolus* (Asteraceae-Heliantheae). Syst Bot Monogr 36: 1–195
Rabakonandrianina E (1980) Infrageneric relationships and the origin of the Hawaiian endemic genus *Lipochaeta* (Compositae). Pac Sci 34: 29–39
Rabakonandrianina E, Carr GD (1981) Intergeneric hybridization, induced polyploidy, and the origin of the Hawaiian endemic *Lipochaeta* from *Wedelia* (Compositae). Am J Bot 68: 206–215
Robichaux RH, Carr GD, Liebman M, Pearcy RW (1990) Adaptive radiation of the Hawaiian silversword alliance (Compositae-Madiinae): ecological, morphological, and physiological diversity. Ann Mo Bot Gard 77: 64–72
Robinson H (1979) Studies in Heliantheae (Asteraceae). XVIII. A new genus *Helianthopsis*. Phytologia 44: 257–269
Robinson H (1981) A revision of the tribal and subtribal limits of the Heliantheae (Asteraceae). Smithson Contrib Bot 51: 1–102
Sanders RW, Stuessy TF, Marticorena C, Silva OM (1987) Phytogeography and evolution of *Dendroseris* and *Robinsonia*, tree-Compositae of the Juan Fernández Islands. Opera Bot 92: 195–215
Schilling EE, Panero JL, Eliasson UH (1994) Evidence from chloroplast DNA restriction site analysis on the relationships of *Scalesia* (Asteraceae: Heliantheae). Am J Bot 81: 248–254
Stuessy TF (1977) Heliantheae – systematic review. In: Heywood VH, Harborne JB, Turner BL (eds) The biology and chemistry of the Compositae. Academic Press, London, pp 621–671
van Balgooy MMJ (1969) A study on the diversity of island floras. Blumea 17: 139–178
van Steenis CGGJ (1973) Woodiness in island flora. Taiwania 18: 45–48
Wagner WL, Herbst DR, Sohmer SH (1990) Manual of the flowering plants of Hawaii, 2 vols. University of Hawaii Press, Honolulu
Watson EE (1929) Contributions to a monograph of the genus *Helianthus*. Pap Mich Acad Sci Arts Lett 9: 305–475

5 Vertebrate Patterns on Islands

J. ROUGHGARDEN

5.1 Introduction

Vertebrates from islands have motivated the foundation of evolutionary biology and ecology as we know it today. Darwin visited the Galapagos islands during his voyage on the HMS Beagle and encountered the finches, now named after him, that provoked his thoughts on evolutionary change. These finches suggested to Darwin that species were not immutable types, in contrast to the pure elements and compounds of chemistry. Pure water is the same everywhere. Each continent and island does not have its own form of water, but each continent and island may very well contain a unique and indigenous form of finch, or cactus, and so forth, implying that biological types are not universal and are subject to modification over evolutionary time. Thus, from its beginning, the scientific investigation of vertebrates on islands has emphasized evolutionary distinctness, and even today studies of ecological processes on islands include evolutionary processes as well.

The contributions of islands to contemporary ecology also began with the Galápagos. In the 1940s, Lack observed that the body sizes of Darwin's finches varied among islands in a way that reflected competition between species. The idea is that species with different body (and beak) sizes consume different resources, and thereby partition the total array of resources in a way that lowers interspecific competition and permits coexistence (Lack 1947). This idea dominated research on island vertebrates during the next four decades. In the 1960s, patterns that related island size and distance from the mainland to bird species diversity were discovered by MacArthur and Wilson. Their work spawned the subfield of biogeography called island biogeography, that continues to have great importance today, especially for conservation biology, wherein islands are viewed as prototypes of habitat fragments. Thus, the vertebrates on islands have been important since the beginning of modern evolutionary biology and ecology, and many studies of island vertebrates are found in standard textbooks on evolution and ecology. This chapter offers some contemporary perspectives on the classic patterns of island vertebrates, and mentions some of the new findings that extend the earlier work in new directions.

Department of Biological Sciences and Department of Geophysics, Stanford University, Stanford, California 94305, USA

Ecological Studies, Vol. 115
Vitousek et al. (eds) Islands

5.2 Islands and Biodiversity

The basic pattern of biodiversity on islands relates both the area of an island and its distance from a source area (usually the adjacent mainland) to the number of species on it. Specifically, if the island is sufficiently close to the source area, then the number of species on it increases as the island's area to some exponent. For islands that are not close to the source area, the ratio of the number on it to the number expected on it given its area if it were close to the source area decreases as a linear function of the distance from the source area. These patterns are called the area and distance effects of island biogeography (Diamond 1973).

MacArthur and Wilson proposed in 1963 that the fauna on any island is the steady state (or equilibrium) level of biodiversity set by a balance of immigration and extinction on the island. They visualized an island as continually receiving propagules of new species, with the rate of arrival of new species to the island being a decreasing function of the number of species already on the island. They also supposed that the total rate at which species are becoming extinct on the island increases with the number of species on it. Graphically, the immigration rate of new propagules to the island can be drawn as a decreasing function of the number on it, and the extinction rate as an increasing function, so that the intersection of these curves represents the steady state biodiversity on the island on the horizontal axis, and the steady state turnover rate on the vertical axis. This hypothesis is known as the "turnover" theory of island biogeography.

By itself, the hypothesis that the biodiversity on an island is continually turning over, while maintaining a constant total biodiversity, says nothing about the area and distance effects of biogeography. To address these biogeographic patterns, MacArthur and Wilson added to the turnover model two additional assumptions: that the immigration rate to an island decreases with its distance from the source area, and that the extinction rate on an island increases as the island becomes smaller.

Thus, the classic theory of island biogeography (MacArthur and Wilson 1967) is today identified with three kinds of propositions, the area and distance effects that are solely descriptions of empirical patterns, the hypothesis that the biodiversity on islands is in a continual state of turnover, and the further hypothesis that the area and distance effects are explained by the biogeographic variation of immigration and extinction rates in relation to island area and distance from the source fauna.

While island biogeography theory has been valuable in conservation biology, few conservation biologists are aware of important academic debate about it. Conservation biologists tend to accept all of island biogeography theory as an integral whole, not realizing that some parts may be correct and others incorrect for various taxa. However recent findings in island biogeography affect conservation strategies and conservation biologists should not content themselves with knowledge that is one to two decades out of date.

The main issue of academic debate in island biogeography concerns the reality of turnover. Is biodiversity on islands in a continual state of turnover? The alternative is that an island "fills up" to a level that depends on its area and variety of habitats, and that thereafter any propagules to the island fail because no ecological space is left. Williamson in 1983 termed turnover "ecologically trivial" in birds on the island to Skokholm because the core species were unaffected while the propagules going extinct never had an opportunity to establish because the island was already full.

Most textbooks draw the immigration curve $I(S)$, as a linear decreasing function of the number of species on the island, S, and the extinction rate $E(S)$ as a linear increasing function of S. In fact, in 1981 Gilpin and Diamond measured the extinction and immigration rates of birds in the Solomon Islands as convex curves: $I(S)$ varies as $(1 - S/P)^7$ and $E(S)$ varies as $S^{2.5}$ where P is number in source fauna. Thus, the immigration rate declines more slowly than linear and the extinction rate rises faster than linear, a shape called convex, and is illustrated as sharply scalloped curves. The implication is that the islands tend to have a core fauna, with very high extinction rates and very low addition rates for any species beyond this core. While Gilpin and Diamond view the situation as still consistent with the turnover model, with the qualification that turnover is not uniformly distributed throughout the fauna, Williamson interprets essentially the same situation as consistent with the qualitatively different model wherein species in the core have no turnover and species outside the core fail at the beach head.

Still further evidence against turnover in vertebrates comes from the observation of "nested subset species-area relationships". This pattern is one wherein the smallest islands have a certain species, say A, then the next larger has that plus another, i.e., A and B, a still larger island has A, B, and C, and so forth. Each island's fauna is a proper subset of a larger island's fauna until the largest island is reached. I have illustrated this pattern for the herpetofauna of the British Virgin Islands based on data from Lazell in 1983, and it is also known for lizards in the Bahamas (cf. Roughgarden 1994). This pattern is not consistent with the turnover model. According to the turnover model, each island small enough to have only one species from the source area should have a random species from the source pool of species (Puerto Rico for the Virgin Islands), and each island with two species should have a random pair of species.

Thus, the existence of a relation between biodiversity and island area does not itself imply that turnover is taking place as well. The Virgin Islands definitely show an increasing species-area relation, but the nested subset pattern of species identities rules out the turnover theory of island biogeography as an explanation for biodiversity in this system of islands.

For vertebrates, as has been seen, there is substantial reservation about the reality of turnover in the biodiversity of islands; but for invertebrates the data for turnover remain strong and convincing. In 1969–70, Simberloff and Wilson experimentally defaunated the arthropods on small mangrove islands in the Florida cays. They directly observed both the approach to a stable steady-state biodiversity as these mangrove islands were recolonized by arthropods and the continual turnover of species within the islands even after the steady-state biodiversity was attained. The extinction of arthropod species within the mangrove islands was balanced by the arrival of new species that took their place. More recently, in the Bahamas, Schoener in 1983 directly quantified the relative propensity or arthropods and vertebrates to exhibit turnover and showed that, indeed, insects exhibit a higher turnover than lizards.

The implications for conservation seem quite clear. To conserve vertebrates large islands are needed. Small islands tend to be homogeneous, and even a great number of them will not contain much biodiversity. I should note, too, that the comparative energetics of the various classes of vertebrates affect conservation strategies. Homeotherms such as birds and mammals are probably more susceptible both to natural and human-induced extinction than poikilotherms such as snakes and lizards because insectivorous lizards can be as much as 100 times more abundant than insectivorous birds on the same food supply.

5.3 Islands as Microcosms

Islands are valuable in the study of ecological processes because they have fewer species on them relative to similar physical situations on the continents. The relatively simplified communities on islands invite the prospect that ecological principles can be discovered more readily on islands than on continents.

Still, how properly to compare an island with a continent is an open question because of qualitative differences between them. First, an insular vertebrate fauna typically differs from its continental counterpart in having fewer large predators, so that a given taxon functions higher in the tropic web on islands than on continents. Small lizards themselves become top predators in the absense of the birds that usually prey on them, but they remain different enough from both continental lizards and from birds of prey to hinder comparisons. Second, vertebrates tend to assume sizes on islands that they do not have on the continent. Small animals, such as rodents, tend to become larger, and large animals tend to become smaller, such as pigmy elephants and deer. The functionality of such insular forms may not compare with continental forms. Third, the landscape scale is largely absent on islands, with local diversity less reflective of regional diversity than on continents. There may be other differences as well that are suspected but poorly documented. Perhaps insectivorous vertebrates are more generalized in the sense of consuming more prey types on islands than on continents because their prey are less likely to have specialized chemical defenses. All of these differences conspire to make islands unique in their ecology, and to prevent their being used as miniatures, or scale models, of continental ecosystems. Of course, one might view island ecology as important in its own right, which is certainly true, but it remains a difficult issue to explain exactly which findings from islands can be transferred to continental situations.

For example, islands have been used for over four decades as model systems for studying interspecific competition. In Fig. 17 of *Darwin's Finches* Lack observed in 1947 that two islands each have a solitary ground-finch with a bill depth of 9–10 mm. On one of these islands the solitary species is *Geospiza fortis* and on the other island is it *G. fuliginosa*; but on three islands these same two species coexist with each other, and there *G. fuliginosa* has a bill depth of about 8 mm while *G. fortis* has a bill depth of about 12–13 mm. Thus, each species has diverged in opposite directions from the solitary size when they occur with one another. This pattern invited the interpretation that on the two-species islands two solitary-sized species have evolutionarily diverged in body size from one another when they came in contact, a phenomenon termed "character displacement" by Brown and Wilson (1956).

These ideas were pursued further with *Anolis* lizards in the Lesser Antilles of the eastern Caribbean (cf. Roughgarden 1994). Over 16 islands have a solitary species with a body length of about 65 mm, and 8 islands have two species, with the smaller of these less than the solitary size, and the larger of these larger than the solitary size. Experimental studies confirmed that the species compete for food and that the strength of interspecific competition is stronger the more similar the body sizes of the competitors. Competition is almost surely involved in explaining the biogeographic pattern of body sizes in *Anolis*. Two scenarios for how competition can cause the body size pattern are presently favored by various authors. The first is the original idea of divergent evolution of two initially similar species. The second postulates that the two species now living together on a two-species island acquired body size differences while in allopatry in their ancestral habitats. These differences then allowed each to invade the other's range. Either way competition is involved, but in one scenario the differences that permit coexistence where evolved in sympatry and in the other in allopatry.

Perhaps the most surprising finding in the competition studies with *Anolis* is that the community composition does not appear to be explainable with ecological principles alone. The communities appear to be undersaturated and their composition therefore reflects the transport processes that have carried the species to these islands. In the eastern Caribbean these processes are mostly plate tectonics combined with short-range over-water dispersal. The premise of community studies during the 1960s and 1970s was that species interactions determined who would live with whom, and that dispersal and speciation were somehow fast enough to introduce the populations appropriate to fill any ecological vacancies. Instead, the membership in communities reflects the vagaries of the applicant pool more than the wishes of the existing members.

While this all makes for an interesting chapter in the history of ecological ideas, the question remains, what can we learn about continents from islands? I feel that the relation of an island to a continent is analogous to that between a tissue culture and an organism. An island is an extraction from a continent, often left alone for a long time. It reveals how a special subset of the processes in a continental ecosystem work, but its functioning cannot be extrapolated to the whole. To use islands well, we must play to their strengths, study them for a deeper understanding of the processes that do occur there, and not view them as continental miniatures.

References

Brown WL, Wilson EO (1956) Character displacement. Syst Zoo 5: 49–64

Diamond J (1973) Distributional ecology of New Guinea birds. Science 179: 759–769

Gilpin M, Diamond J (1981) Immigration and extinction probabilities for individual species: relation to incidence functions and species colonization curves. PNAS 78: 392–396

Lack D (1974) Darwin's finches. Cambridge University Press, Cambridge

Lazell JD Jr (1983) Biogeography of the herpetofauna of the British Virgin Islands, with description of a new anole (Sauria: Iguanidae). In: Rhodin AGJ, Miyata K (eds) Advances in herpetology and evolutionary biology. Museum of Comparative Zoology, Cambridge MA, pp 99–117

MacArthur RH, Wilson EO (1963) An equilibrium theory of insular zoogeography. Evolution 17: 373–387

MacArthur RH, Wilson EO (1967) The theory of island biogeography. Princeton University Press, Princeton

Roughgarden JD (1994) *Anolis* lizards of the Caribbean: ecology, evolution and plate tectonics. Oxford University Press, Oxford

Schoener T (1983) Rate of species turnover decreases from lower to higher organisms: a review of data. Oikos 41: 372–377

Simberloff D, Wilson EO (1969) Experimental zoogeography of islands: the colonization of empty islands. Ecology 50: 278–296

Simberloff D, Wilson EO (1970) Experimental zoogeography of islands: a two year record of colonization. Ecology 51: 934–937

Williamson M (1983) The land-bird community of Skokholm: ordination and turnover Oikos 41: 378–384

6 Patterns of Diversity in Island Soil Fauna: Detecting Functional Redundancy

D. FOOTE

6.1 Introduction

Distinguishing between functional redundancy and complexity is critical to developing an understanding of the consequences of species loss in ecosystems. An important case is species-rich communities closely associated with ecosystem function. It is unclear whether high species diversity reflects an increased number of functional groups or simply a higher degree of redundancy. Soil organisms represent one example of a community that affects ecosystem processes and is highly diverse. In these terrestrial detritivore-decomposer systems, the breakdown of litter and the subsequent mineralization of nutrients are carried out by a community that can approach 1000 species per m^2 in some temperate woodland soils (Anderson 1978). This chapter explores a method to distinguish between functional complexity and redundancy in soil organisms using comparisons of biogeographic patterns of species diversity in communities of island soil fauna.

The analysis of community structure across environmental gradients frequently reveals changes in dominance within taxa or guilds (Whittaker 1967, Odum 1971). In this sense, a guild (Root 1967) may be equated with a functional group, where species share attributes of potential significance to a particular ecosystem-level process (Vitousek and Hooper 1993). The distribution of members of a functional group along an environmental gradient where biogeochemical processes vary provides a preliminary test of functional redundancy. If one defines functional groups in terms of their members having similar effects on specific ecosystem processes, then change in species occurrence within the membership of a given group over such a gradient implies lack of functional redundancy: even if two species have the same effect on a given ecosystem-level biogeochemical process, if they respond to the environment in a manner different from one another they cannot be considered equivalent. Alternatively, if two species in a functional group coexist across the same gradient, they may be viewed as potentially redundant, at least in the context of a specific type of environmental change represented by the gradient. In the case of multiple species, lack of functional redundancy along a gradient would be demonstrated by increased beta diversity within the functional group.

Cooperative Park Resources Study Unit, National Biological Survey, P. O. Box 52, Hawaii National Park, Hawaii 96718, USA

Ecological Studies, Vol. 115
Vitousek et al. (eds) Islands

In practice, patterns of diversity within a functional group along an environmental gradient will be influenced by other physical variables, differences in dispersal abilities among functional group members, and species interactions both within the group and with members of the community at large. The strength of inferences regarding species sensitivity to specific environmental factors can be increased by choosing simple gradients. High tropical islands are attractive in this regard because they offer often steep gradients of temperature and precipitation on a small spatial scale (Carlquist 1980). Volcanic islands are especially useful because transects along environmental gradients can frequently be located on even-aged lava flows or ash deposits, producing a constant substrate. Age gradients of long-term soil development can also be sampled across different-aged flows or volcanoes or archipelagos. These features are best represented in the Hawaiian Islands, which have been used as a model system to study ecosystem function (Vitousek and Benning, Chap. 7, this Vol.).

The identification of biogeographic patterns of diversity in species-rich taxa of soil organisms is primarily impeded by lack of adequate taxonomic information. The systematics of microorganisms and soil fauna are poorly developed or incomplete for most groups in many localities (Eversham et al. 1992; Hawksworth 1992). In addition to providing simple independent environmental gradients, isolated oceanic islands also display taxonomically more manageable patterns of species diversity. Diversity is much lower than on larger land masses and island fauna are disharmonic, resulting from relatively few successful colonizers giving rise to an extensive radiation of closely related species (Zimmerman 1970; Williamson 1981). Overall, fewer systematists can accomplish a great deal towards describing the total faunal diversity.

This chapter reviews the role of soil fauna in ecosystem processes, the taxonomic impediments to working with such a diverse fauna and the use of islands as a model system for the analysis of biogeographic patterns of diversity across environmental gradients. Most of the discussion focuses on the Hawaiian Islands because certain groups of soil fauna there have been thoroughly characterized taxonomically (Nishida 1992), the biological organization of soil fauna communities along an elevational gradient has been described (Mueller-Dombois et al. 1981), and the structure and function of Hawaiian montane forest ecosystems has received considerable attention recently (Vitousek et al. 1988, 1992; Vitousek and Benning, Chap. 7, this Vol.). The utility of using biogeographic comparisons of species diversity to distinguish between complexity and redundancy in functional groups is addressed using data from the literature on two well-known taxa of Hawaiian soil fauna, *Drosophila* (pomace flies) and Collembola (springtails).

6.2 The Role of Soil Fauna in Ecosystem Processes

The International Biological Program, including the Origin and Structure of Ecosystems Program, in the early 1970s supported the first intensive research into the role of soil invertebrates in ecosystem processes in many biomes. The primary observation in terrestrial systems has been that 80% or more of nutrients and energy bound up in primary producers is ultimately returned to the soil as dead organic matter (Swift et al. 1979). Research on the role of detritivores and other soil fauna in decomposition and nutrient mineralization of litter has continued as part of a broad program outlined by Pomeroy (1970) to study ecosystem processes through analysis of nutrient cycling. The comminution of leaf litter by soil fauna followed by the mineralization or immobilization of nutrients by microbes are the pathways by which decomposition and nutrient fluxes occur in terrestrial ecosystems. The direct contribution of soil invertebrates to decomposition is minor. However, even though the invertebrate fauna accounts for less than 10% of the total detritus-decomposer respiration, many microbivores and detritivores have been shown to play a large indirect regulatory role in decomposition through feeding activities and the modification of soil structure during burrowing (Anderson et al. 1981; Petersen and Luxton 1982; Seastedt 1984).

The effect of diversity on ecosystem processes has largely focused on the description of detritus food webs and the categorization of functional groups in both natural and agroecosystems. Functional groupings have traditionally been made using body size and major taxonomic divisions (Table 6.1). Body size largely determines if an animal is capable of altering soil and litter structure (e.g., earthworms) or is constrained by it (e.g., springtails) (Anderson 1988). Other groupings have tended to focus on ecological roles, but there is an implicit assumption that these groupings correspond to specific effects on ecosystem processes (Hendrix et al. 1986; Lee 1991). Intensive research on population biology of soil organisms in some terrestrial ecosystems has allowed for greater precision in grouping soil fauna according to trophic behavior, including the proportion of different foods ingested, mode of feeding, and distribution in the soil profile (Anderson 1975; Moore et al. 1988). The major groupings outlined in Table 6.1 suggest broad functional groups. Where information is available on individual species population biology, finer groupings will be possible.

Table 6.1. Soil invertebrates classified by size

Group	Size (mm)	Examples
Microfauna	<2	Protozoa
		Nematodes
Mesofauna	2–10	Collembola (springtails)
		Acari (oribatid mites)
Macrofauna	10–20	Diptera (flies)
		Coleoptera (beetles)
		Hymenoptera (ants)
Megafauna	>20	Annelids (earthworms)
		Diplopoda (millipedes)

6.3 Taxonomic Impediments

The identification of functional groups among soil organisms and the description of biogeographic trends in species diversity within groups is complicated by a lack of taxonomic information for many taxa. Soil fauna of many forest and grassland ecosystems, as well as agroecosystems, have been studied intensively (Lohm and Persson 1977; Mitchell and Nakas 1986; Swift and Anderson 1993). However, the total species diversity of soil organisms has not been described for any habitat or region (Lee 1991). This is the consequence of the high diversity of soil fauna combined with poor taxonomic information for many groups. In comparing species richness of ants in Australia to North Amercia, Morton (1993, p 165) states that "the alpha taxonomy of Australian ants is in a woeful state, largely because the sheer diversity of the fauna has daunted systematists". The same may be said for many other soil invertebrates, particularly within the microfauna and mesofauna. For example, in a typical square meter of continental temperate forest soil, oribatid mites and Collembola comprise about 90% of the mesofauna, but there are usually in excess of a 100 species present among just these two groups. Species diversities of these taxa are in the thousands across major continents: it has been estimated that the approximately 700 described species of Nearctic Collembola represent 70–75% of the total species diversity (Richards 1979; Waltz 1990); similarly, Greenslade (1991) estimates that there are close to 2000 species of Collembola in Australia, including many undescribed forms. The more diverse soil mites represent a larger challenge; as few as 6% of species in the infraorders comprising the North American oribatid mites have valid names, based on a recent estimate that the total fauna comprises 15000 species (O'Connor 1990).

Some island ecosystems offer detailed taxonomic information on soil fauna for biogeographic comparisons. For example, many soil taxa from large developed island groups, such as Britain, Japan, and New Zealand have been described (Rapport 1982). However, patterns of species diversity among most other archipelagos are generally poorly known, with the exception of prominent litter macrofauna, such as tiger beetles (Freitag 1992; Pearson and Juliano, 1993), ground beetles (Darlington 1970; Liebherr 1988; Moore 1992; Niemalä 1992), and certain groups of ants (Wilson 1959; Goldstein 1975). In these cases, there are often good regional descriptions of diversity. Knowledge of island detritivore and microbivore communities is typically poor, but intensive research has been carried out on some groups from several archipelagos. For example, Collembola have been surveyed recently on the Canary Islands (Fjellberg 1992) and in Hawaii, where the last of 15 volumes published in the *Insects of Hawaii* series is devoted to springtails (Christiansen and Bellinger 1992).

High, typically volcanic, oceanic islands display a rich range of different taxa. However, the species diversity is manageable at a taxonomic level. For example, the number of described Acari in Hawaii (687, Nishida 1992) is only about one-tenth that of North American species (5106, O'Connor 1990). Likewise, there are 172 named species of Collembola in the archipelago (Nishida 1992).

Christiansen (pers. comm.) estimates that this number represents approximately 92% of the Hawaiian springtail fauna. Table 6.2 shows the improvement in the archipelago's taxonomic knowledge of springtails reflected in successive volumes of the *Insects of Hawaii*. A similar volume for a continent such as Australia would have to encompass approximately 2000 species (Greenslade 1991).

6.4 Patterns of Diversity for Island Soil Fauna in Hawaii

Functional groups of soil organisms in Hawaii illustrate the potential for using gradient analysis to reveal different levels of community complexity and potential functional redundancy in ecosystems. Hawaii provides a model system where gradients are steep, often simple, and provide the isolation and spatial scale where species diversity is rich, yet taxonomically manageable. Vitousek (1995) has recently reviewed the patterns of variation in Hawaiian ecosystems. Five major state factors that affect ecosystem function are climate, time, organisms, relief and parent material. Continuous, well-defined gradients of temperature, precipitation, and time occur in this archipelago. This is a consequence of: (1) a temporal sequence of volcanic activity that is stationary under the northwestward moving Pacific plate, forming new volcanic islands at the southeast end of the chain (MacDonald et al. 1983); (2) The position of the Hawaiian islands approximately perpendicular to prevailing northeasterly tradewinds; and (3) the latitude (19° to 28° N) where sea level temperatures are warm and cool rapidly with increasing elevation up the slopes of the high volcanoes (Carlquist 1980). In contrast, the remaining state factors change little. Structural dominants in many vegetation types have broad distributions (cf. Gagne and Cuddihy 1990) and the shield-building phases of the volcanoes produce areas with constant relief and similar volcanic parent material (MacDonald et al. 1983). As a consequence, simple independent climatic and age gradients occur over the islands, where vegetation, relief, and parent material remain relatively constant (Vitousek 1995). Furthermore, as a consequence of anthropogenic changes in Hawaii, and the introduction of nonindigenous species, there are also gradients of disturbance to soils in Hawaii, such as changes in the frequency of digging by feral pigs.

Drosophila and Collembola are taxonomically two of the best known groups of Hawaiian soil fauna. These insect taxa also represent dominant components of

Table 6.2. Removal of a taxonomic impediment: numbers of described genera and species of Collembola from two sequential volumes of the Insects of Hawaii

Reference	Genera	Species	Endemic sp.	Est. % described
Zimmerman (1948)	24	32	0?	?
Christiansen and Bellinger (1992)	49	154	92	92

two functional groups. Springtails are mesofauna that act primarily as grazers of fungi in litter and soil, and contribute to the regulation of microbial decomposer populations (Moore et al. 1988). Collembola display ametabolous development, such that larvae differ from adults primarily in size and occupy the same habitat. *Drosophila* are macrofauna that share with Collembola the role of microbivores. However, larvae undergo complete metamorphosis and adult flies occupy different microhabitats from larvae. *Drosophila* larvae complete their development primarily in decaying leaves, fruit and fungi. A number of species also appear to be specialized to feed on microorganisms associated with the early stages in the decomposition of bark and these *Drosophila* often colonize rotting bark prior to branch fall. Within the genus, Hawaiian *Drosophila* are saprophytic on more than 40 families of native plants and the majority of species appear to be specialized to breed on a single host plant family (Carson and Kaneshiro 1976). Larval *Drosophila* ingest decaying plant material together with microbes so they may be considered detritivores as well as microbivores. *Drosophila* presumably share with Collembola a regulatory function in the decomposition of decaying plant tissue and also contribute to the comminution of litter through their feeding activities.

The Collembola fauna in Hawaii is comprised of 172 species, of which 96 are endemic to the islands, 75 are adventive (meaning immigrant species that were not purposely introduced), and one species is of unknown status (Nishida 1992). The genus *Drosophila* in Hawaii includes 355 described species (Carson 1992) of which only 15 are adventive. The species diversity within Collembola is spread relatively evenly over 50 genera, with two exceptions: the genera *Entomobrya* and *Lepidocyrtus* contain 25 and 16 species, respectively (Christiansen and Bellinger 1992). The large numbers of *Drosophila* reflect a dramatic case of evolutionary radiation of species within the genus in Hawaii, to the extent that Hawaiian *Drosophila* make up approximately one-fifth of the *Drosophila* fauna worldwide. In addition to representing two separate functional groups, Collembola and *Drosophila* therefore differ in total diversity, the degree of endemism, and the magnitude of diversity within genera. The distribution of species across the archipelago is known from published records for both groups (Nishida 1992) and their distribution along an altitudinal transect has been described (Mueller-Dombois et al. 1981). Biogeographic patterns in species diversity of the two taxa along three separate gradients are examined. Because the island biota is composed of both endemic and adventive species and the latter are often cosmopolitan species with broad geographical ranges, adventive and endemic species are analyzed separately within each taxa to evaluate how the degree of functional redundancy might vary both between and within taxa of soil organisms.

6.4.1 Age Gradient

The Hawaiian Islands form a geochronosequence: among the major islands, in the northwest the island of Kauai is approximately 5.1 ma in age. At the southeast end of the chain, Mauna Loa Volcano on the island of Hawaii is approximately

0.5 ma old and on its flank the current eruption of Kilauea Volcano is actively creating new land (MacDonald et al. 1983; Clague and Dalrymple 1987). Patterns of island endemism in *Drosophila* and Collembola can be compared along this gradient. The potential for functional redundancy within each taxa is expressed by the number of islands over which a given species is found: if all species within a functional group occupied all islands there would be the greatest opportunity for redundancy. Alternatively, loss of functional redundancy will occur as a consequence of changes in species composition within guilds on different islands and this wil be expressed by increased levels of single island endemism among the taxa.

Figure 6.1 shows the frequency distribution of species of endemic and adventive *Drosophila* and Collembola that occur on one to four of the major islands or island group. Close to 90% of the endemic *Drosophila* are single-island endemics, compared to less than 50% of the endemic Collembola. At the other extreme, nearly 20% of the endemic Collembola are distributed across the entire archipelago, whereas only 1% of the endemic *Drosophila* have this wide a distribution. The distribution of endemic Collembola suggests a substantial degree of potential functional redundancy: more than one-third of the species of springtails occur over three or more islands. The opportunity for functional redundancy is also generally greater among adventive species than among endemic species. However, the difference is greatest within the *Drosophila*. The distributional trends of

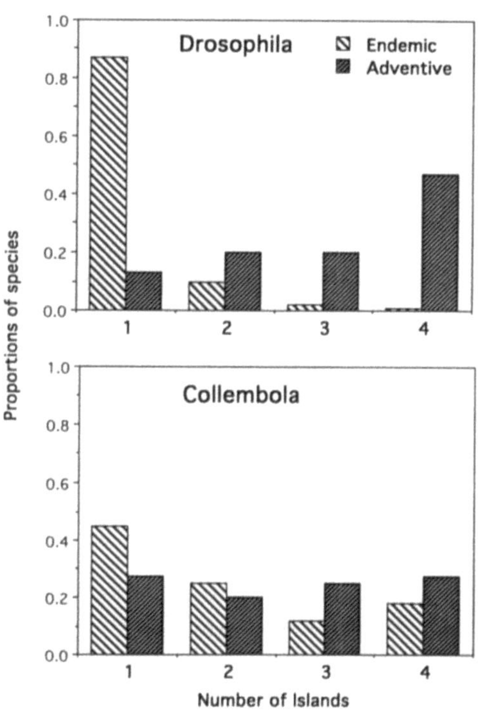

Fig. 6.1. The distribution of endemic and adventive *Drosphila* and Collembola on the four major Hawaiian islands or island group, expressed as the proportion of species within a group occurring on one to four islands. The island group is Maui Nui, that includes the islands of Maui, Molokai, and Lanai, all of which were joined in the recent geological past. (MacDonald et al. 1983)

adventive *Drosophila* is opposite to that observed in endemic *Drosophila*. More than 40% of the adventive fly species have been recorded on all four islands.

A complication interpreting island distributions of the two taxa is that separate taxonomic treatments of allopatric populations may involve different weightings of characters that define species limits (cf. Ridley 1986). Also, it is usually the practice to treat similar allopatric populations as conspecific unless there is good evidence to the contrary (Mayr 1969). The systematics of Hawaiian *Drosophila* have been studied more intensively than Collembola and, as a consequence, there will tend to be a bias towards recognizing more species among similar allopatric populations in the better-studied group. In the present comparison there is good evidence that the differences in species diversity are real. Hawaiian *Drosophila* exhibit great diversity in secondary sexual characters (Kaneshiro 1983) lacking in Collembola and there are complex courtship patterns that appear to make populations of Hawaiian *Drosophila* more prone to speciation events (Kaneshiro 1988).

The observation that adventive species in both taxa tend to have broader distributions suggests that dispersal ability plays a role in creating opportunities for functional redundancy within these taxa. One possibility is that the frequency of inter-island dispersal of Collembola and adventive *Drosophila* has been increased by human activity on the islands. Endemic Collembola appear to exhibit overall much lower levels of habitat specialization compared to endemic *Drosophila*. The native host plant associations of the latter may have hindered opportunities for movement, whereas soil and fruit associated with food plants have been moved between the Hawaiian Islands since the arrival of Polynesians more than 1500 years ago.

Because substrate age varies both within and between islands and the existing biogeographic data identifies only island records of species occurrence, determination of taxa sensitivity to substrate age must await further survey work. The sensitivity of these soil taxa to substrate age can be better tested on the smaller geographical scale of a single island, where substrates at otherwise environmentally similar sites can vary by as much as 150000 years (Riley and Vitousek 1995).

6.4.2 Elevational Gradient

Comparisons of biogeographic patterns of diversity in Hawaiian *Drosophila* and Collembola can also be made over an elevational gradient in montane forests on the east slope of Mauna Loa volcano on the island of Hawaii. Soil and litter arthropods were sampled as part of the Hawaii International Biological Program on Island Ecosystems that supported research on Mauna Loa from 1971 to 1976. Soil fauna was sampled from 12 sites between 1220 and 2440 m on the Mauna Loa Transect. The vegetation ranged from rainforest at the low elevation site to subalpine forest at the highest three locations; all the remaining sites occurred in a montane seasonal zone. The transect crossed two old deep ash deposits that caused variation in substrate age and composition, so the gradient is complex in this instance. The primary environmental gradient was a change in mean annual air temperature that ranged from 6 °C in the alpine section to 16 °C in the montane sections of the transect (Mueller-Dombois et al. 1981).

The soil fauna was sampled over an approximately 2 year period. Collembola were collected using pitfall traps and Berlese extraction of soil and litter samples. Three genera of drosophilids (*Drosophila*, *Idiomyia*, and *Scaptomyza*) were collected using banana baits, and reared from litter samples. The numbers of individuals collected were not reported in a manner that allowed comparisons of relative abundance between the two taxa, so only the presence or absence of a given species is reported. When some of the surveys described below were performed, more than 70% of the soil microarthropod species collected were new records for Hawaii (Mueller-Dombois et al. 1981). The comparisons that follow are possible because of the improvement in taxonomic knowledge of *Drosophila* (Hardy and Kaneshiro 1981) and Collembola (Christiansen and Bellinger 1992; Table 6.2).

The distribution of drosophilids and Collembola over the sites is shown in Table 6.3 and suggests the same differences as in response of the two groups to the age gradient. Endemic drosophilids were only present at an average of approximately two sites each, while adventive Collembola occurred at a mean of over eight sites. The observation that endemic Collembola on the age gradient produced a pattern similar to adventive species on both gradients suggests that gradient sensitivity in this taxa is not driven by degree of endemism. The potential for functional redundancy may be more of an inherent feature of given taxa, such as Collembola, within functional groups.

The five species of adventive *Drosophila* were present on an average of more than six sites, and the contrast between endemic and adventive drosophilids indicates that whether or not one is an adventive species makes a difference for this taxa in terms of potential redundancy. There is also some evidence of gradient sensitivity among the adventive species of *Drosophila*. The number of adventives present along the transect increases with temperature. However, there is also considerable evidence of habitat specialization limiting the distributions of *Drosophila* along the transect. Most of the endemic species are limited to areas in or near sites 3 and 4. These are patches of forests (called kipukas) growing on older deep ash substrates surrounded by a younger lava flow. At least two endemic drosophilids, *D. mimica* and *I. engyochracea*, complete their life cycles on a species of tree, *Sapindus saponaria*, that is restricted to these kipukas along the transect (Carson and Kaneshiro 1976). The adventive species may also be limited by habitat preferences. For example, the species with the greatest altitudinal range, *D. simulans*, has also been shown to be the most generalized in choice of breeding substrate in comparison to *D. buskii*, *D. hydei*, and *D. immigrans* in Australia (Oakeshott et al. 1982). Therefore, the effect of elevational gradients on many of these drosophilid populations is most likely mediated indirectly through the effect of changes in vegetation on habitat preferences. Increased habitat specialization along an environmental gradient, like reduced dispersal ability, will tend to reduce potential functional redundancy within a taxa.

Table 6.3. Presence or absence of drosophilid flies and Collembola along the Mauna Loa Transect[a]

Species	Elevation site	1220 m 1	2	3	4	5	6	7	8	9	10	11	2400 m 12	
Drosophilids														Status[b]
Drosophila mimica			−	+	+		−	−	−	−	−	−	−	END
D. imparisetae			−	+	+		+	−	−	−	−	−	−	END
D. reducta			−	+	+		+	−	−	−	−	−	−	END
D. carnosa			−	−	+		−	−	−	−	−	−	−	END
D. chaetopeza			−	−	+		−	−	−	−	−	−	−	END
Idiomyia fungiperda			−	+	−		+	−	−	−	−	−	−	END
I. engyochracea			−	+	+		−	−	−	−	−	−	−	END
Scaptomyzaparalobae			−	+	+		+	+	+	−	−	−	−	END
S. cuspidata			−	−	+		+	+	+	−	−	−	−	END
S. near cuspidata			−	−	+		−	−	−	−	−	−	−	END
S. articulata			−	−	−		+	−	−	−	−	−	−	END
S. inaequalis			−	−	−		−	−	−	+	−	−	−	END
D. hydei			+	+	−		+	−	−	−	−	−	−	ADV
D. busckii			+	+	+		+	+	+	−	−	−	−	ADV
D. sulfurigaster			+	+	+		+	+	+	−	−	−	−	ADV
D. immigrans			+	+	+		+	+	+	+	+	−	−	ADV
D. simulans			+	+	+		+	+	+	+	+	+	+	ADV
Collembola														
Anurophorus lohi		−		+	+	+	+		−	+	+	+	+	END
Sminthurides lolelua		−		+	−	−	−		−	−	−	−	−	END
Entomobrya multifasciata		−		+	+	+	+		+	+	+	+	+	ADV
E. atrocincta		−		+	+	+	+		+	+	+	+	+	ADV
E. socia		+		+	+	+	+		+	+	+	+	+	ADV
E. sauteri		+		+	+	+	+		+	+	+	+	+	ADV?
E. purpurascens		−		+	−	+	+		−	+	+	+	+	ADV
Lepidocyrtus ruber		+		+	+	+	+		+	+	+	−	+	ADV
Salina celebensis		+		+	+	+	+		+	+	+	−	+	ADV
Brachytomella parvula		+		+	+	+	+		+	+	+	+	+	ADV
Hypogastrura boletivora		+		+	+	−	+		+	+	+	−	+	ADV
Onychiurus folsomi		+		+	+	+	+		+	+	+	−	+	ADV
O. encarpatus		+		+	+	+	+		+	+	+	−	+	ADV
Neanura muscorum		+		+	−	−	+		−	+	+	−	−	ADV
Sminthurinus elegans		−		+	+	−	+		+	−	−	+	−	ADV

[a] A blank column indicates taxa was not sampled at that site.
[b] Status: END = endemic, ADV = adventive.

6.4.3 Disturbance Gradients

Environmental gradients reflect changes in ecosystem state factors. In contrast, biological invasions often provide gradients of disturbance in native Hawaiian ecosystems. The precise nature and regime of disturbance will be easiest to characterize where individual nonindigenous species alter ecosystem function. For example, two particularly invasive species with ecosystem-level effects in Hawaiian forests are the faya tree (*Myrica faya*) and feral pigs (*Sus scrofa*) (Vitousek 1986; Cuddihy and Stone 1990). The faya tree is an introduced actinorrhizal nitrogen-fixer that can dramatically increase inputs of fixed nitrogen into ecosystems developing on recent volcanic substrates that are naturally low in nitrogen (Vitousek et al 1987; Vitousek and Walker 1989). In contrast, the digging and rooting activities of feral pigs in forest soils has the potential to increase decomposition of leaf litter and the rate at which nitrogen and other nutrients become available and subsequently lost from the soil through leaching (Singer et al. 1984; Vitousek 1986). Feral pigs also alter the composition of vegetation and can enhance the growth of invasive alien plants (Diong 1983; Stone et al. 1992).

Vtorov (1993) studied the distribution of Collembola in three fenced feral pig exclosures in the Olaa Forest of Hawaii Volcanoes National Park that ranged in age (and presumably recovery from pig disturbance) from 0 to 7 years. The distribution of Collembola across this temporal gradient of exclosure age was similar to the distributions over the environmental gradients discussed above: among 16 species (either adventive or of unknown status) close to one-third were found at all three sites. However, of the species restricted to a single site, all were found in the oldest exclosures. Foote and Carson (1994) examined changes in relative frequencies of 14 species of endemic *Drosophila* in the youngest and oldest exclosure from 1971 to 1993. There was a decline in relative frequency of members of a guild of endemic *Drosophila* that bred on native Lobelioids, a group sensitive to pig disturbance. Despite the construction of exclosures, two of the four species of lobelioid breeders are now missing from Olaa Forest. No differences in the numbers of endemic or adventive *Drosophila* species were observed between the two sites (Foote, unpubl.).

The ecosystem-level effects of pigs have not been measured at these sites. However, it appears that both Collembola and *Drosophila* are sensitive to pig disturbance. More work needs to be done to examine the spatial scale over which pig disturbance may influence potential functional redundancy in these taxa and the relative influence of changes in host plant community structure and long-term soil fertility.

6.5 Conclusion

Soil fauna represents a diverse biological system where one can examine evidence for functional redundancy within and between functional groups associated with specific ecosystem-level processes. Analyzing biogeographic patterns of diversity in functional groups across environmental gradients provides one method to identify species that appear potentially redundant. Islands provide a model system where patterns of diversity can be explored because species diversity is present at a taxonomically manageable level, and high tropical volcanic islands, in particular, offer a rich range of simple, independent environmental gradients over which changes in diversity can be measured. Patterns of biological diversity in Collembola and *Drosophila* over both age and elevational gradients in the Hawaiian Islands reveal differences in potential functional redundancy both within and between these groups. Regardless of the level of endemism, Collembola display a much higher degree of species overlap across both gradients than do endemic *Drosophila*, indicating that the former taxa possess greater functional redundancy in the context of these two gradients. The observation that adventive *Drosophila* display much lower gradient sensitivity compared to endemic members of the genus suggests that the magnitude of potential functional redundancy in soil organisms will vary both between and within taxa of different functional groups.

It may be possible to examine the long-term ecosystem consequences of species loss within functional groups caused by particular regimes of disturbance, such as feral pig activity in Hawaii. Data on *Drosophila* populations from Olaa Forest in Hawaii Volcanoes National Park suggest that one consequence of prolonged pig disturbance may be the loss of endemic species. The result will be increases in the relative frequency of adventive *Drosophila* and an associated increase in the level of functional redundancy within this taxa.

Acknowledgments. Carla D'Antonio, Tom Dudley, Karin Schlappa and Peter Vitousek commented critically on an earlier version of this manuscript.

References

Anderson JM (1975) Succession, diversity and trophic relationships of some soil animals in decomposing leaf litter. J Anim Ecol 44: 475–495

Anderson JM (1978) Inter- and intra-habitat relationships between woodland *Cryptostigmata* species diversity and diversity of soil and litter microhabitats. Oecologia 32: 341–348

Anderson JM (1988) Fauna-mediated transport processes in soils. Agric Ecosyst Environ 24: 5–19

Anderson RV, Coleman DC, Cole CV (1981) Effects of saprotrophic grazing on net mineralization. In: Clark FE, Roswall T (eds) Terrestrial nitrogen cycles. Processes, ecosystem strategies and management impacts. Ecol Bull (Stockholm) 33: 201–216

Carlquist SJ (1980) Hawaii, a natural history. Pacific Tropical Botanical Garden, Lawai, Kauai, Hawaii

Carson HL (1992) Inversions in Hawaiian *Drosophila*. In: Krimbas CB, Powell JR (eds) *Drosophila* inversion polymorphism. CRC Press, Ann Arbor, pp 407–439

Carson HL, Kaneshiro KY (1976) *Drosophila* of Hawaii: systematics and ecological genetics. Annu Rev Ecol Syst 7: 311–345

Christiansen K, Bellinger P (1992) Insects of Hawaii vol 15 collembola. University of Hawaii Press, Honolulu

Clague DA, Dalrymple GB (1987) The Hawaiian-Emperor volcanic chain. In: Decker RW, Wright TL, Stauffer PH (eds) Volcanism in Hawaii. US Geological Survery Professional Paper 1350. US Government Printing Office, Washington, DC, pp 5–84
Cuddihy LW, Stone CP (1990) Alteration of native Hawaiian vegetation. Cooperative National Park Resources Studies Unit, Honolulu
Darlington PJ Jr (1970) Insects of Micronesia, Coleoptera: Carabidae including Cicindelinae. BP Bishop Mus Insects Micronesia 15: 1–49
Diong CH (1983) Population ecology and management of the feral pig (*Sus scrofa* L.) in Kipahulu Valley, Maui. Ph D Dissertation University of Hawaii, Honolulu
Eversham BC, Jolliffe AS, Davis BNK (1992) Soil macrofauna. In: Groombridge B (ed) Global biodiversity: status of the earth's living resources. Chapman & Hall, London, pp 103–115
Fjellberg A (1992) Collembola of the Canary Islands. I. Introduction and survey of the family Hypogastruridae. Entomol Scand 22: 437–456
Foote D, Carson HL (1994) *Drosophila* species as monitors of change in Hawaiian ecosystems. National Biological Survey Status and Trends Source Book. National Biological Survey, Washington, DC
Freitag R (1992) Biogeography of West Indian tiger beetles (Coleoptera: Cicinidelidae). In: Noonan GR, Ball GE, Stork NE (eds) The biogeography of ground beetles of mountains and islands. Intercept, Andover, UK, pp 123–158
Gagne WC, Cuddihy LW (1990) Vegetation. In: Wagner WL, Herbst DR, Sohmer SH (eds) Manual of the flowering plants of Hawai'i vol 1. University of Hawaii Press, Honolulu, pp 45–114
Goldstein EL (1975) Island biogeography of ants. Evolution 29: 750–762
Greenslade PJ (1991) Collembola (springtails). In: The insects of Australia: a textbook for students and research workers. 2nd edn. CSIRO, Cornell University Press, Ithaca
Hardy DE, Kaneshiro KY (1981) Drosophilidae of Pacific Oceania. In: Ashburner M, Carson HL, Thomson JN Jr (eds) Genetics and biology of *Drosophila* vol 3A Academic Press, London, pp 309–348
Hawksworth DL (1992) Microorganisms. In: Groombridge B (ed) Global biodiversity: status of the earth's living resources. Chapman & Hall, London, pp 47–54
Hendrix PF, Parmelee RW, Crossley DA Jr, Coleman DC, Odum EP, Groffman PM (1986) Detrius food webs in conventional and no-tillage agroecosystems. BioScience 36: 374–380
Kaneshiro KY (1983) Sexual selection and direction of evolution in the biosystematics of Hawaiian Drosophilidae. Annu Rev Entomol 28: 161–178
Kaneshiro KY (1988) Speciation in the Hawaiian *Drosophila*: sexual selection appears to play an important role. Bioscience 38: 258–263
Lee KE (1991) The diversity of soil organisms. In: Hawksworth DL (ed) The biodiversity of microorganisms and invertebrates: its role in sustainable agriculture. CAB International, Wallingford
Liebherr JK (1988) Biogeographic patterns of West Indian *Platynus* carabid beetles (Coleoptera). In: Liebherr JK (ed) Zoogeography of Caribbean insects. Cornell University Press, Ithaca, pp 121–152
Lohm U, Persson T (eds) (1977) Soil organisms as components of ecosystems. Proc 6th Int Soil Zool Ecol Bull 25, Stockholm
MacDonald GA, Abbott AT, Peterson FL (1983) Volcanoes in the sea 2nd edn. University of Hawaii Press, Honolulu
Mayr E (1969) Principles of systematic zoology. McGraw-Hill, New York, pp 193–197
Mitchell MJ, Nakas JP (eds) (1986) Microfloral and faunal interactions in natural and agro-ecosystems. Developments in biogeochemistry. Nijhoff/Junk, Dordrecht
Moore BP (1992) The Carabidae of Lord Howe Island (Coleoptera: Carabidae). In: Noonan GR, Ball GE, Stork NE (eds) The biogeography of ground beetles of mountains and islands. Intercept, Andover, UK, pp 159–173
Moore JC, Walter DE, Hunt HW (1988) Arthropod regulation of micro- and mesobiota in below-ground detrital food webs. Annu Rev Entomol 33: 419–439
Morton SR (1993) Determinants of diversity in animal communities of arid Australia. In: Ricklefs RE, Schluter D (eds) Species diversity in ecological communities. Historical and geographical perspectives. University of Chicago Press, Chicago, pp 159–169

Mueller-Dombois D, Bridges KW, Carson HL (eds) (1981) Island ecosystems: biological organization in selected Hawaiian communities. Dowden Hutchinson Ross, Stroudsburg, PA
Niemelä J (1992) Distribution of carabid beetles in the Aland archipelago, SW Finland (Coleoptera: Carabidae). In: Noonan GR, Ball GE, Stork NE (eds) The biogeography of ground beetles of mountains and islands. Intercept, Andover, UK, pp 175–187
Nishida G (ed) (1992) Hawaii terrestrial arthropod checklist. BP Bishop Museum, Honolulu
Oakeshott JG, May TW, Gibson JB, Willcocks (1982) Resource partitioning in five domestic *Drosophila* species and its relationship to ethanol metabolism. Aust J Zool 30: 547–556
O'Connor BM (1990) The North American Acari: current status and future projections. In: Kosztarab M, Schaefer CW (eds) Systematics of the North American insects and arachnids: status and needs. Virginia Agric Exp Sta Info Ser 90–1. Virginia State University, Blacksburg
Odum EP (1971) Fundamentals of ecology, 3rd edn. WB Sanders, Philadelphia, PA
Pearson DL, Juliano SA (1993) Evidence for the influence of historical processes in cooccurrence and diversity of tiger beetle species. In: Ricklefs RE, Schluter D (eds) Species diversity in ecological communities. Historical and geographical perspectives. University of Chicago Press, Chicago, pp 194–202
Petersen H, Luxton M (1982) A comparative analysis of soil fauna populations and their role in decomposition processes. Oikos 39: 287–388
Pomeroy LR (1970) The strategy of mineral cycling in ecosystems. Annu Rev Ecol Syst 1: 171–190
Rapoport EH (1982) Areography. Pergamon Press, Oxford
Richards WR (1979) Collembola. In: Dank HV (ed) Canada and its insect fauna. Mem Entomol Soc Canada No 108
Ridley M (1986) Evolution and classification: the reformation of cladism. Longman, London
Riley RH, Vitousek PM (1995) Nutrient dynamics and trace gas flux during ecosystem development in Hawaiian montane rainforest. Ecology (in press)
Root R (1967) The niche exploitation pattern of the blue-grey gnatcatcher. Ecol Monogr 37: 317–350
Seastedt TR (1984) The role of microarthropods in decomposition and mineralization processes. Annu Rev Entomol 29: 25–46
Singer FJ, Swank WT, Clebsch EEC (1984) Effects of wild pig rooting in a deciduous forest. J Wildl Manage 48: 464–473
Stone CP, Cuddihy LW, Tunison JT (1992) Responses of Hawaiian ecosystems to removal of feral pigs and goats. In: Stone CP, Smith CW, Tunison JT (eds) Alient plant invasions in native ecosystems of Hawaii: management and research. University of Hawaii Press, Honolulu, pp 666–704
Swift MJ, Anderson JM (1993) Biodiversity and ecosystem function in agricultural systems. In: Schulze D, Mooney HA (eds) Biological diversity and ecosystem function. Springer, Berlin Heidelberg New York
Swift MJ, Heal OW, Anderson JM (1979) Decomposition in terrestrial ecosystems. Blackwell Scientific Publications, Oxford
Vitousek PM (1986) Biological invasions of ecosystem properties: can species make a difference? In: Mooney HA, Drake J (eds) Biological invasions of North America and Hawaii. Springer, Berlin Heidelberg New York, pp 163–176
Vitousek PM (1995) The Hawaiian Islands as a model system for ecosystem studies. Pacific Science 49: 2–16
Vitousek PM, Benning TL (1994) Ecosystem and landscape diversity: islands as model systems. Biological diversity and ecosystem function on islands. Springer, Berlin Heidelberg New York
Vitousek PM, Hooper D (1993) biological diversity and terrestrial ecosystem biogeochemistry. In: Schulze D, Mooney HA (eds) Biological diversity and ecosystem function. Springer, Berlin Heidelberg New York
Vitousek PM, Walker LR (1989) Biological invasion by Myrica faya in Hawaii: plant demography, nitrogen fixation, and ecosystem effects. Ecol Monogr 59: 247–265

Vitousek PM, Walker LR, Whiteaker LD, Mueller-Dombois D, Matson PA (1987) Biological invasion by *Myrica faya* alters ecosystem development in Hawaii. Science 238: 802–804
Vitousek PM, Matson PA, Turner DR (1988) Elevational and age gradients in Hawaiian montane rainforest: foliar and soil nutrients. Oecologia 77: 565–570
Vitousek PM, Aplet G, Turner DR, Lockwood JJ (1992) The Mauna Loa environmental matrix: foliar and soil nutrients. Oecologia 89: 372–382
Vtorov IP (1993) Feral pig removal: effects on soil microarthropods in a Hawaiian rain forest. J Wildl Manage 57: 875–880
Waltz RD (1990) Order Collembola. In: Kosztarab M, Schaefer CW (eds) Systematics of the North American insects and arachnids: status and needs. Virginia Agric Exp Sta Info Ser 90–1, Virginia State University, Blacksburg
Whittaker RH (1967) Gradient analysis of vegetation. Biol Rev 49: 207–264
Williamson M (1981) Island populations. Oxford University Press, Oxford
Wilson EO (1959) Adaptive shift and dispersal in a tropical ant fauna of Melanesia. Evolution 13: 122–144
Zimmerman EC (1948) Insects of Hawaii, vol. 1. Introduction. University of Hawaii Press, Honolulu, 206 pp
Zimmerman EC (1970) Adaptive radiation in Hawaii with special reference to insects. Bio-tropica 2: 32–38

7 Ecosystem and Landscape Diversity: Islands as Model Systems

P. M. VITOUSEK and T. L. BENNING

7.1 Introduction

Most of the effort that has gone into relating biological diversity and ecosystem function – in this volume and elsewhere – has focused on biological diversity at the species level. That is where most information is available, and that is where the modern epidemic of species invasions and extinctions on islands has altered diversity most spectacularly.

Biological diversity at higher and lower levels of biological organization also can affect the functioning of ecosystems, which we define (rather narrowly) as: the fluxes of energy, water, and elements in an area; exchanges of energy, or water or nutrients between terrestrial ecosystems and the atmosphere or aquatic systems; and/or changes in state (soil development and fertility, disturbance frequency and intensity) of an area. Kaneshiro (Chap. 3, this Vol.) describes the nature and significance of genetic- and population-level diversity; in this chapter, we are concerned with the effects of landscape-level biological diversity on ecosystem function.

Analyses of the causes and consequences of heterogeneity at the level of ecosystems and landscapes have received substantial attention recently (cf. Turner 1989; Kolasa and Pickett 1991; Turner and Gardner 1991) as part of the reemergence of landscape ecology as a science concerned with addressing linkages between landscape-level structure and function (Forman and Godron 1986). The definition of landscape patterns, the characterization of ecological processes that affect those patterns, and the dynamics of pattern and process over time are central to landscape ecology (cf. Urban et al. 1987; Swanson et al. 1988; Burke 1989; Dunn et al. 1991), and an examination of the functional significance of diversity at the ecosystem or landscape level would be a natural extension of this approach.

Diversity at the ecosystem level is caused by a differential response of species and/or ecosystem functions to variation (both temporal and spatial) in environmental conditions such as topography, drainage, parent material, and natural and anthropogenic disturbance. Ecosystem-level diversity contributes to the development and maintenance of regional diversity at the species and population levels; it has long been known that "habitat heterogeneity" is an excellent predictor of species diversity, in part because β-diversity makes a substantial

Department of Biological Sciences, Stanford University, Stanford, California 94305, USA

Ecological Studies, Vol. 115
Vitousek et al. (eds) Islands

contribution to overall species diversity. Mechanistically, a high degree of spatial or temporal heterogeneity interferes with competitive displacement and consequently allows the persistence of high species diversity (Harris 1986). Moreover, ecotones (transition zones between ecosystems), edges (boundaries between ecosystems), and habitat mosaics are essential to the success of numerous species. It is also clear that arthropogenic modifications of ecosystem diversity can alter population and species diversity (cf. Turner 1989; Peters and Lovejoy 1992), most often by reducing or eliminating specialized species and those that require large contiguous areas of habitat.

Can diversity at the ecosystem or landscape level affect ecosystem function more directly? Can the overall magnitude of energy, water, or element cycling and/or flux differ in a landscape that supports a diversity of ecosystems, in comparison with an otherwise-equivalent homogeneous landscape? There is evidence that landscape diversity can have important effects, and that those effects can be additive or interactive. Additive effects are easily understood, if not always easily determined. For example, fluxes of the radiatively active trace gas nitrous oxide from tropical forests vary as a function of soil fertility and land use, and regional estimates of its flux must account for this ecosystem-level diversity. Matson et al. (1990) used a combination of satellite remote sensing-based ecosystem classifications and on-the-ground flux estimates to calculate that while pastures covered less than 12% of a $> 800\ km^2$ area north of Manaus, Brazil, they accounted for more than 40% of nitrous oxide emission from the region.

Interactive effects of ecosystem-level diversity are more dynamic and interesting. For example, Peterjohn and Correll (1984) examined the importance of riparian forests in an agricultural landscape in Maryland, USA. Fertilizer nitrogen is applied to upland portions of the watershed; a portion of it leaches through the soil and moves through groundwater and interflow towards streams. En route, the nitrogen moves through intact riparian forests, where some is taken up into aggrading biomass and/or denitrified to nitrogen gas. The presence of ecosystem-level diversity in this agricultural landscape thereby affects both overall losses of nitrogen from the region, and the pathways by which nitrogen is lost.

Other examples of interactive effects of ecosystem-level diversity on ecosystem function have been described; they include studies of ecological processes along soil catenas in grassland ecosystems (cf. Schimel et al. 1985, 1991), analyses of interactions between N emission from agricultural ecosystems, its deposition on natural ecosystems, and consequent changes in their diversity and functioning (Arnolds 1991; Bobbink 1991; Berendse et al. 1993), and an elegant analysis of nutrient cycling, production, and integrated ecosystem dynamics along a permafrost-constrained flowpath in arctic Alaska (Giblin et al. 1991).

In this chapter, we ask if there is anything unique or particularly useful about island ecosystems that can help us to evaluate interactions between ecosystem-level diversity and ecosystem function. Can research on islands provide information on the nature and causes of ecosystem-level diversity? On the interaction between diversity at the ecosystem and species/population levels? On the direct effects of ecosystem diversity on ecosystem function? We conclude that island ecosystems in general, and volcanic island arcs in particular, are useful for analyzing the controls of ecosystem-level diversity, much as islands have been useful as model systems for understanding evolutionary processes and ecosystem dynamics. Islands often contain an extraordinary amount of ecosystem diversity in a small geographical area, and the causes of that diversity are relatively well defined and constrained. To the extent that a clear understanding of the causes of ecosystem diversity is important to understanding its effects on ecosystem function, islands will also be useful in evaluating ecosystem diversity/ecosystem function interactions.

7.2 Ecosystem Diversity on Islands

Islands are valuable for studying the causes and consequences of biological diversity at the ecosystem level for many of the same reasons that they are valuable for studies of evolution – the same processes are important on islands as on continents, but they operate against a simpler and clearer background on islands.

Following the soil scientist Hans Jenny (1941, 1980), we think of ecosystem structure and function as being controlled by a number of "state factors", of which the most important are climate, organisms, relief, parent material, time, and (increasingly) human activity. These factors are not always the proximate regulators of ecosystem function; rather, they represent ultimate controls. For example, the productivity of many temperate forest ecosystems is limited by nitrogen availability. Nitrogen availability is not a state factor, but it is controlled (ultimately) by interactions of the major state factors. In simple cases, the influence of a particular state factor can be examined by locating situations in which one of the factors varies while all of the others remain more or less constant. More often, however, the state factors interact with each other, and there may be several levels of controls and feedbacks between these ultimate controls and ecosystem function. These complex situations (which make up most of the earth) are most easily addressed through the use of ecosystem models and experiments (cf. Parton et al. 1987). Even so, models and experimental approaches are best developed in sites where the interactive effects of multiple state factors are minimal or well controlled.

The structure and functioning of ecosystems on islands as well as continents vary in response to variations in the major state factors. A number of biological, geological, and climatic features interact to cause spatial variation in some of these factors to occur at a much finer scale on islands than continents, resulting in more finely divided landscape patterns. Moreover, some of the state factors are more constant on islands than is ever the case in continental areas, thus making it more straightforward to determine what factors are important in controlling ecosystem-level diversity. In this section, we will show how variation in state factors controls ecosystem-level diversity on islands. We emphasize examples from the Hawaiian Islands, but will make use of examples from other islands and archipelagoes where appropriate.

7.3 The State Factors and Their Interactions

7.3.1 Climate

Climate is the single most important factor controlling ecosystem structure and function; indeed, temperature and precipitation individually probably account for more variation in ecosystems than any single other factor. By definition, islands experience a maritime climate, with relatively small seasonal variations in temperature. However, temperature decreases with increasing elevation, generally at an environmental lapse rate of 5.5 – 6 °C/1000 m, and even tropical islands can thereby support very cold climates.

Variation in precipitation arises from montane and rain shadow effects. For example, the east and northeast sides of the island of Hawaii experience very high rainfall because they face into prevailing northeast trade winds, while the leeward side of smaller mountains in dry (Fig. 7.1). Large islands also can affect precipitation by altering local atmospheric circulation – the southwest side of the island of Hawaii is sheltered from the tradewinds, and it would seem that it should be dry, but the mass of the island and the overhead summer sun are sufficient to establish a land-sea breeze cycle, and lead to substantial summer rain (Fig. 7.1).

As a consequence of variation in temperature and precipitation, even a single small island like Hawaii can encompass most of the range in variation of tropical climates (*Altas of Hawaii* 1983). This fine scale of variation in climate is not unique to islands; continental regions can support similarly strong climatic gradients, although the much-discussed altitudinal compression of vegetation zones on isolated mountains (cf. Grubb 1977; Bruijnzeel et al. 1993) is particularly well developed on islands. As discussed below, it is more the lack of variation in other factors that makes climatically caused variation in island ecosystems of interest.

7.3.2 Organisms

It is a truism that plants affect soils and soils affect plants in a positive feed-back – or, for those interested in the ultimate controls on ecosystem function, a vicious circle (Jenny 1980). Jenny dealt with this problem by considering the regional flora and fauna – the organisms that could occupy a site – as the independent factor driving soil formation and ecosystem structure and function.

The organism factor is of interest on islands for two reasons. First, due to their isolation, oceanic islands generally support a much lower diversity of species than continental areas, and much of that diversity can be traced to evolutionary radiation from a small number of initial colonists. Consequently, some island species (or clusters of close relatives) cover an extraordinarily broad range of environmental conditions, potentially making the organism factor much more constrained on oceanic islands than continents. This process reaches an extreme in the Hawaiian Islands, where the dominant native tree *Metrosideros polymorpha* covers sites from sea level to treeline, occurs as the first woody colonist on young

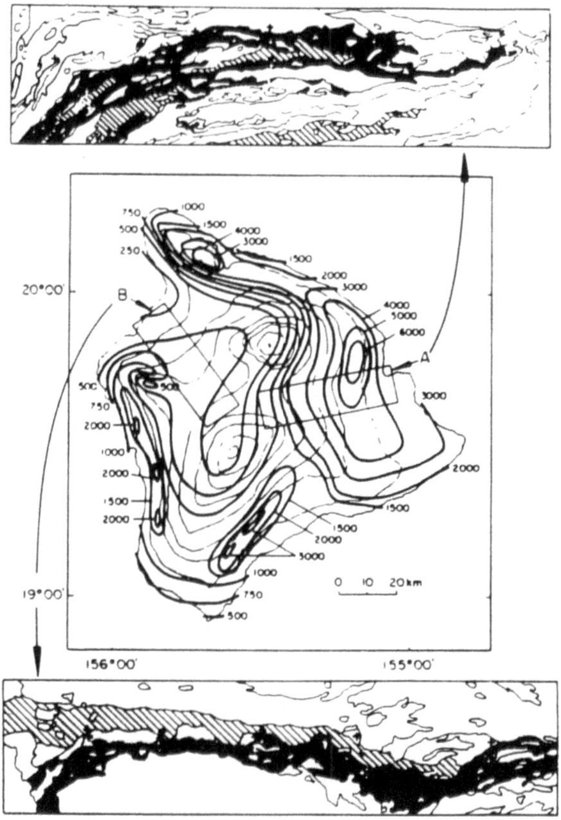

Fig. 7.1. Variation in climate on the Island of Hawaii. Temperature varies with elevation (*fine lines* represent 500-m contours); precipitation (*heavy lines*, in mm/a) varies with elevation, aspect, and wind patterns. Rainfall is high year-round on the east side of the island, and very low most of the year on the northwest. The southeast side receives rain from winter storms, while the southwest receives summer rain due to a land-sea breeze cycle. (After Giambelluca et al. 1986; Atlas of Hawaii 1983). The rectangular-Sections (*A* and *B*) illustrate lava flows of different ages and textures that cross a wide range of elevations on the wet and dry flanks. From Vitousek et al. (1992)

lava flows and on the oldest soils in the islands, and dominates a rainfall gradient from the wettest sites on earth (> 10 000 mm/a) to sites receiving < 400 mm/a (although only early in primary succession in the driest sites) (Dawson and Stemmerman 1990; Stemmerman and Ihsle 1993). No continental species, and few others on islands, dominate over such a broad range of environments.

Islands also are useful for studies of the organism factor because biological invasions by exotic species are notoriously successful on islands (D'Antonio and Dudley; Chap. 9, this Vol.), and these invasions represent a fundamental shift in the "organism factor". The consequences of this change for ecosystem function can therefore be determined directly on islands more readily than on continents (Vitousek and Walker 1989; Hughes et al. 1991; D'Antonio and Vitousek 1992).

7.3.3 Relief

Relief (or topography) affects ecosystem structure and function because it represents both a cause and a consequence of resource redistribution across landscapes. For example, water and soil are removed from hillslopes and deposited on foot-slopes and alluvial areas, thereby affecting ecosystem function in both donor and recipient area (cf. Schimel et al. 1985).

Island ecosystems can contain a wealth of topographic variation that affects ecosystem function (cf. Mueller-Dombois 1988; Bowden et al. 1992), just as do continental ecosystems. To the extent that islands can be used to examine variation in ecosystems caused by relief while keeping other factors more or less constant, island ecosystems can be valuable for studies of this factor. More interestingly, one particular type of island comes as close to lacking relief as any set of ecosystems on Earth, and so offers the opportunity to keep this factor constant while varying others. Young shield volcanoes in island arcs derived from the movement of tectonic plates over convective plumes in the mantle ("hot spots") are built up by eruptions of relatively fluid lavas; they have a gentle angle of repose; they are porous so that water erosion is minimal; and their surface is renewed frequently by new lava flows (MacDonald et al. 1983; Clague and Dalrymph 1987; Decker et al. 1987). For example, Mauna Loa Volcano in Hawaii is 4168 m tall but has a smooth, gentle, virtually unbroken surface (save for rough lava, and occasional fault scarps and eruptive vents) over most of its 5500 km^2 surface, an area that covers a very broad range of climates (Fig. 7.1). Older shield volcanoes in the Hawaiian Islands (and elsewhere) have experienced substantial erosion, and their relief can be used to make an important contribution to understanding ecosystem structure and function in these areas (Mueller-Dombois 1988).

7.3.4 Parent Material

Parent material is the substrate from which soils are derived, and in which ecosystems develop; it can exert an important control on soil fertility, water budgets, and ecosystem structure and function. This factor is particularly interesting on volcanic islands because of the lack of variation in the chemistry of parent material. In Hawaii, the lavas erupted during the shield-building stage of volcanism are all tholeiitic basalt or its close relatives (Wright and Helz 1987); individual eruptions differ in their magnesium mixing ratio but in little else of ecological interest, in comparison with the wide range of rock types to be found in most continental regions. Consequently, it is possible to hold the chemistry of parent material virtually constant across a wide range of climates or a long time of soil development.

While the chemistry of parent material varies relatively little, the texture of parent material on volcanic islands can vary substantially. Parent material in Hawaii ranges from extensive areas of tephra deposited from lava fountains or rare explosive eruptions, to lava flows of pahoehoe and aa lava. (Pahoehoe flows are less viscous than aa: they have a smooth, ropy surface when they cool. Aa is rough and broken, with a solid core and many individual rocks on the surface: MacDonald et al. 1983).

The relative simplicity of the parent material in Hawaii and other volcanic archipelagoes does not extend to all islands; many contain a diversity of parent material no different from that found in diverse continental areas. For example, Puerto Rico supports volcanic rock, limestone, and serpentine all in close proximity, and all distributed along a strong moisture gradient (Medina et al. 1994); New Caledonia supports a similarly wide variety of parent material (Mueller-Dombois, Chap. 13, this Vol.). To the extent that other state factors can be kept constant, the ecosystem diversity caused by variation in parent material could be addressed profitably in these systems.

7.3.5 Time

The time over which an area has been exposed to rock weathering, leaching, and the activity of organisms determines the structure and fertility of soils, and hence the functioning of ecosystems, to a substantial extent (Smeck 1985; Vitousek and Walker 1987; Walker and Syers 1976). Volcanic islands, especially hot spot-derived island arcs, are uniquely valuable systems for determining the ecosystem-level consequences of the time factor.

The Hawaiian Islands are the best known and most dynamic example of a hot spot-derived archipelago, although the Galápagos, most of the South Pacific high island chains, and the Mascarenes have been formed similarly. Hawaii results from the northwesterly movement of the Pacific tectonic plate over a stationary convective plume in the mantle; there have been islands present in that location for at least 70 million years. The hot spot is now located under the southeastern edge of the island chain; it supports frequent eruptions by the active volcanoes Mauna Loa and Kilauea. Historical flows from these volcanoes are well mapped, and most of the prehistoric flows have been mapped and dated using ^{14}C from buried organic matter derived from plants covered by these flows (Lockwood and Lipman 1980; Lockwood et al. 1988). These maps yield a high-resolution base of information on the location and distribution of a wide range of substrates ranging in age up to several thousand years old. For older substrates, the volcanoes to the northwest on Mauna Loa and Kilauea on the island of Hawaii and on the older islands to the northwest make up a well-defined and dated sequence of sites that extends all the way to Kauai and Niihau, at ca. 5 million years the oldest remaining high islands in the Hawaiian chain (MacDonald et al. 1983; Decker et al. 1987; Riley and Vitousek 1995).

7.3.6 Interactions

With several colleagues, we are evaluating how the climate and time factors interact to control ecosystem structure and function in the Hawaiian Islands. We use the unique features of young, hot spot-derived shield volcanoes to hold organisms, relief, and parent material chemistry relatively constant as described above. Climate and time are the remaining major variables, and their influence can be determined relatively straightforwardly because most of the diversity in ecosystems is organized into discrete lava flows (Lockwood et al. 1988). Each of the several hundred flows on Mauna Loa represents a single-age, single-substrate strip reaching from above the treeline towards, and often to, the sea. Each flow therefore fulfills a field ecologist's fantasy of what a transect should be. Moreover, chronosequences of parallel flows that differ primarily in age have been developed on a given flank of Mauna Loa (Drake and Mueller-Dombois 1993; Aplet and Vitousek 1994); flows of similar age that cross rainforest versus desert can be compared (Vitousek et al. 1990, 1992); and there is even some true replication, with flows of nearly the same age occurring close to each other on the same flank of Mauna Loa.

7.4 Ecosystem Diversity and Ecosystem Function

Given the substantial levels of ecosystem-level diversity observed on many islands, and the unusually clear and comprehensible causes of that diversity, could we use these systems to ask how ecosystem-level diversity affects ecosystem function? We believe that the answer is "yes", but to date little of substance has been done to take advantage of the opportunities that islands present.

The structure of ecosystem diversity on young shield volcanoes in particular seems made to order for determining some of the functional consequences of ecosystem diversity. In those systems, the matrix of lava flows of different ages and textures, all distributed independently along steep gradients of precipitation and temperature, offers extraordinarily fine-scale variation in ecosystem properties for which the causes are known quite well. Moreover, as a result of chance or upslope topography, some otherwise-similar portions of the surface of a shield volcano lack this fine-scale heterogeneity (Fig. 7.2). Do ecosystem functions at a particular site (a flow of a given age at a given elevation) within a complex area (i.e., Fig. 7.2B) differ *because* they are imbedded in a matrix of very different ecosystems? Do any of the major ecosystem functions of the area as a whole (accumulation and/or exchange of energy, water, and elements) differ *because* the area contains a diverse array of ecosystems? If so, does the difference arise *because* of interactions among those ecosystems?

One logical process to examine would be the rate of primary succession, which affects ecosystem- as well as population-level processes through carbon and nutrient accumulation and the development of soils. It is plausible to suggest that any new lava flow might be colonized and develop more rapidly if it occurred within a matrix of successional systems, rather than against a homogeneous background (Fig. 7.2A,B). As a consequence, successional development of the area as a whole might proceed more rapidly in a heterogeneous area than it would if any few components of that area occurred in isolation. This suggestion should

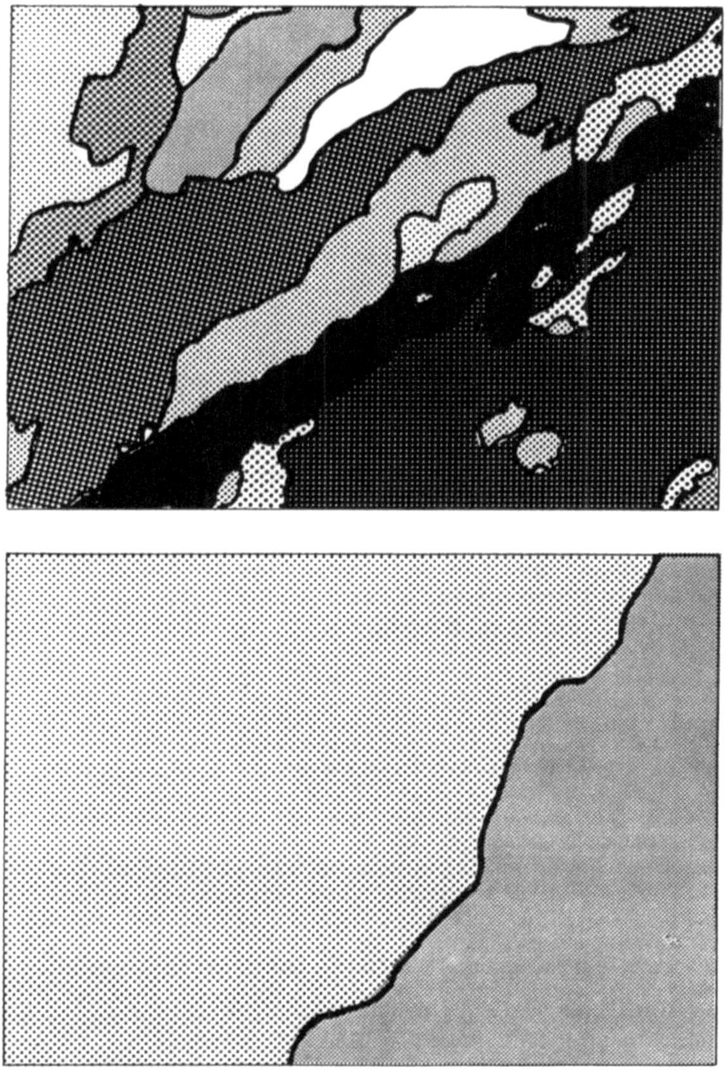

Fig. 7.2 A, B. The matrix of lava flow ages in a relatively homogeneous (*left*) and a heterogeneous (*right*) area on the south side of the northeast rift of Mauna Loa. Relative flow ages are indicated with a gray scale (younger flows are *darker*). The ages of islands of older substrate surrounded by younger flows (kipukas) are indicated similarly. The horizontal dimension of both panels corresponds to 2.5 km. (From the Kipuka Pakekake quadrangle, J. P. Lockwood, pers. comm.)

be testable in Hawaii and elsewhere through careful comparative studies – but it has not been tested. What *has* been observed is that the vegetation on the edge of young lava flows generally is both taller and denser than that in the center of young flows, making it difficult to map lava flows from the air or space (Holcomb 1987). Edge vegetation on young flows can even be better developed than is that on old flows, as a consequence of stand-level dieback on the old flows (cf. Mueller-Dombois 1986) and perhaps other processes. Whether this edge phenomenon represents early colonization due to a seed source in the older ecosystems, microclimate amelioration, greater nutrient availability (because flows are thinner on the their edges, and plants can root into buried soils beneath), complementarity of nutrients supplied by old and young sites, or other processes is not known – but it is amenable to study as to both its causes and its ecosystem-level consequences. Similar situations, in which substantial ecosystem-level diversity occurs for well-known reasons in a small area, occur in other island ecosystems; these could similarly provide a useful resource for determining connections between ecosystem-level diversity and ecosystem function.

Acknowledgments. We thank H. Adsersen and L. Loope for comments on an earlier draft of this manuscript. This work was supported by NSF grant BSR-8918382 to Stanford University and a Pew Foundation Fellowship; TLB was supported by an NSF postdoctoral fellowship.

References

Aplet GH, Vitousek PM (1994) An age-elevation matrix analysis of Hawaiian rainforest succession. J Ecol 82: 137–147

Arnolds E (1991) Decline of ectomycorrhizal fungi in Europe. Agric Ecosyst Environ 35: 209–244

Atlas of Hawaii (1983) University of Hawaii Department of Geography, Honolulu

Berendse F, Aerts R, Bobbink R (1993) Atmospheric nitrogen deposition and its impact on terrestrial ecosystems. In: Vos CC, Opdam P (eds) Landscape ecology of a stressed environment. Chapman & Hall, London, pp 104–121

Bobbink R (1991) Effects of nutrient enrichment in Dutch chalk grassland. J Appl Ecol 28: 28–41

Bowden WB, McDowell WH, Asbury CE, Finley AM (1992) Riparian nitrogen dynamics in two geomorphologically distinct tropical rain forest watersheds. Biogeochemistry 18: 77–99

Bruijnzeel LA, Waterloo MJ, Proctor J, Kuiters AT, Kotterink B (1993) Hydrological observations in montane rain forests on Gunung Silam, Malaysia, with special reference to the 'Massenerhebung' effect. J Ecol 81: 145–167

Burke IC (1989) Control of nitrogen mineralization in a sagebrush steppe landscape. Ecology 70: 115–126

Clague DA, Dalrymple GB (1987) The Hawaiian-Emperor volcanic chain: geologic evolution. In: Decker RW, Wright TC, Stauffer PH (eds) Volcanism in Hawaii. United States Geological Survey, Washington, DC, pp 5–54

D'Antonio CM, Vitousek PM (1992) Biological invasions by exotic grasses, the grass-fire cycle, and global change. Annu Rev Ecol Syst 23: 63–87

Dawson JW, Stemmerman L (1990) *Metrosideros* (Myrtaceae). In: Wagner WL, Herbst DR, Sohmer SH (eds) Manual of the flowering plants of Hawaii. Bernice P. Bishop Museum, Honolulu

Decker RW, Wright TC, Stauffer PH (eds) (1987) Volcanism in Hawaii. United States Geological Survey, Washington, DC
Drake DR, Mueller-Dombois D (1993) Population development of rain forest trees on a chronosequence of Hawaiian lava flows. Ecology 74: 1012–1019
Dunn CP, Sharpe DM, Guntenspergen GR, Stearns F, Yang Z (1991) Methods for analyzing temporal changes in landscape pattern. In: Turner MG, Gardner RH (eds) Quantitative methods in landscape ecology. Springer, Berlin Heidelberg New York, pp 173–198
Forman RTT, Godron M (1986) Landscape ecology. Wiley, New York
Giambelluca TW, Nullet MA, Schroeder TA (1986) Rainfall atlas of Hawaii. Department of Land and Natural Resources, State of Hawaii, Honolulu
Giblin AE, Nadelhoffer KJ, Shaver GR, Laundre JA, McKerrow AJ (1991) Biogeochemical diversity along a riverside toposequence in arctic Alaska. Ecol Monogr 61: 415–435
Grubb PJ (1977) Control of forest growth and distribution on wet tropical mountains, with special reference to mineral nutrition. Annu Rev Ecol Syst 8: 83–107
Harris GP (1986) Phytoplankton ecology: structure, function and fluctuation. Chapman and Hall, New York
Holcomb RT (1987) Eruptive history and long-term behavior of Kilauea Volcano. In: Decker RW, Wright TL, Stauffer PH (eds) Volcanism in Hawaii US Government Printing Office, Washington, DC, pp 261–350
Hughes RF, Vitousek PM, Tunison T (1991) Alien grass invasion and fire in the seasonal submontane zone of Hawaii. Ecology 72: 743–746
Jenny H (1941) Factors of soil formation. McGraw-Hill, New York
Jenny H (1980) Soil genesis with ecological perspectives. Springer, Berlin Heidelberg New York
Kolasa J, Pickett STA (eds) (1991) Ecological heterogeneity. Springer, Berlin Heidelberg New York
Lockwood JP, Lipman PW (1980) Recovery of datable charcoal beneath young lavas: lessons from Hawaii. Bull Volcanol 43: 609–615
Lockwood JP, Lipman PW, Peterson LD, Warshauer FR (1988) Generalized ages of surface lava flows of Mauna Loa Volcano, Hawaii. USGS map Series I-1908 US Government Printing Office, Washington, DC
MacDonald GA, Abbot AT, Peterson FL (1983) Volcanoes in the sea: the geology of Hawaii. University of Hawaii Press, Honolulu
Matson PA, Vitousek PM, Livingston GP, Swanberg NA (1990) Sources of variation in nitrous oxide flux from Amazonian ecosystems. J Geophys Res 95: 16,789–16,798
Medina E, Cuevas E, Figueroa J, Lugo AE (1994) Mineral content of leaves from trees growing on serpentine soils under contrasting rainfall regimes in Puerto Rico. Plant Soil 158: 13–21
Mueller-Dombois D (1986) Perspectives for an etiology of stand-level dieback. Annu Rev Ecol Syst 17: 221–243
Mueller-Dombois D (1988) Vegetation dynamics and slope management on the mountains of the Hawaiian Islands. Environ Conserv 15: 255–260
National Research Council (1993) The role of terrestrial ecosystems in global change. National Academies Press, Washington, DC
Parton WJ, Schimel DS, Cole CV, Ojima DS (1987) Analysis of factors controlling soil organic matter levels in Great Plains grasslands. Soil Sci Soc Am J 51: 1173–1179
Peterjohn WT, Correll DL (1984) Nutrient dynamics in an agricultural watershed: observations on the role of a riparian forest. Ecology 65: 1466–1475
Peters RL, Lovejoy TE (eds) (1992) Global warming and biological diversity. Yale University Press, New Haven, CT
Riley RH, Vitousek PM (1995) Nutrient dynamics and trace gas flux during ecosystem development in Hawaiian montane rain forest. Ecology 76: 292–304
Schimel DS, Stillwell MA, Woodmansee RG (1985) Biogeochemistry of C, N, and P in a soil catena of the shortgrass steppe. Ecology 66: 276–282
Schimel DS, Kittell TGF, Knapp AK, Seastedt TR, Parton WJ, Brown VB (1991) Physiological interactions along resource gradients in a tallgrass prairie. Ecology 72: 672–684
Smeck NE (1985) Phosphorus dynamics in soils and landscapes. Geoderma 36: 185–199

Stemmermann L, Ihsle T (1993) Replacement of *Metrosideros polymorpha*, 'Ōhi'a, in Hawaiian dry forest succession. Biotrop 25: 36–45
Swanson FJ, Kratz TK, Caine N, Woodmansee RG (1988) Landform effects on ecosystem patterns and processes. BioScience 38: 92–98
Turner MG (1989) Landscape ecology: the effect of pattern on process. Annu Rev Ecol Syst 20: 171–198
Turner MG, Gardner RH (eds) (1991) Quantitative methods in landscape ecology. Springer, Berlin Heidelberg New York
Urban DL, O'Neill RV, Shugart HH Jr (1987) Landscape ecology. BioScience 37: 119–127
Vitousek PM, Walker LR (1987) Colonization, succession, and resource availability: ecosystem-level interactions. In: Gray A, Crawley M, Edwards PJ (eds) Colonization, succession, and stability. Blackwell Scientific, Oxford, pp 207–223
Vitousek PM, Walker LR (1989) Biological invason by *Myrica faya* in Hawai'i: plant demography, nitrogen fixation, ecosystem effects. Ecol Monogr 59: 247–265
Vitousek PM, Field CB, Matson PA (1990) Variation in $\delta^{13}C$ in Hawaiian *Metrosideros polymorpha*: a case of internal resistance? Oecologia 84: 362–370
Vitousek PM, Aplet G, Turner D, Lockwood JJ (1992) The Mauna Loa environmental matrix: foliar and soil nutrients. Oecologia 89: 372–382
Walker TW, Syers JK (1976) The fate of phosphorus during pedogenesis. Geoderma 15: 1–19
Wright TL, Helz RT (1987) Recent advances in Hawaiian petrology and geochemistry. In: Decker RW, Wright TC, Stauffer PH (eds) Volcanism in Hawaii. USGS Professional Paper 1350, Washington, DC, pp 625–640

Section B
Threats to Diversity on Islands

8 Prehistoric Extinctions and Ecological Changes on Oceanic Islands

H.F. JAMES

8.1 Introduction

"In addition to the absence of native land birds over large areas in Hawaii, there is another interesting, as well as frustrating, feature about the distribution of the birds: some ecological niches are virtually devoid of native birds.

No native land bird inhabits the forest floor. The long-extinct rail on Hawaii did so, but almost nothing is known about either the habits or the ecological distribution of this small bird."

A. Berger (1972: p. 24)

The body of ecological theory that relates specifically to islands was developed mainly through inference from modern patterns of diversity. Observations of process are limited, because they are usually restricted to the relatively short and recent periods of time covered by most field studies and experiments. Whether the processes that are important in these contexts can be extrapolated to explain biogeographic patterns that have been thousands to millions of years in genesis, is an essential question for which there are few relevant data. In this chapter, I argue that greater reliance on paleontological data would strengthen our understanding of island ecosystems, by linking modern patterns to processes that occur on evolutionary timescales.

The first step in this direction is to recognize the limitations to the information that can be obtained. Paleontological data have inherent biases, and are usually not directly comparable to ecological data. For example, fossil assemblages may be subject to time-averaging, or the mixing of diachronic populations in a single sample. Ecologists can pose research questions in terms that call for data collected with fine temporal resolution. To examine the same phenomena paleontologically, the questions have to be recast in terms that are tolerant of time-averaging. The discussion of turnover rates in this chapter is an example of such a problem.

The inability to observe organisms in life, interacting with their environment, is another obstacle to ecological interpretations of paleontological evidence. To an extent, this loss of autecological detail can be compensated for by reconstructing parameters of ecological importance, such as the weight, diet, and

Department of Vertebrate Zoology, National Museum of Natural History, Washington, DC 20560, USA

Ecological Studies, Vol. 115
Vitousek et al. (eds) Islands

paleohabitat of extinct animals (e.g., Atkinson and Millener 1991). In this chapter, the absence of autecological detail makes it necessary to use a very coarse-grained approach when grouping extinct Hawaiian birds into feeding guilds.

Turning to the island fossil record itself, the salient observation that has emerged from many years of research is of an extinction event that significantly reduced vertebrate species richness during the postglacial period (the Holocene epoch, 10000 years ago to the present). Within the island realm, these extinctions were global in scope. They can be traced through the Pacific, Atlantic, Arctic, and Indian Oceans, and the Mediterranean and Caribbean Seas. A recent review lists over 200 species of extinct island birds known from fossils (Milberg and Tyrberg 1993), a number that is steadily increasing with new discoveries. Many other endemic island vertebrates were lost, such as tortoises, lizards, bats, giant rodents, and dwarf ungulates.

Quaternary extinctions on continents are variously attributed to climatic reversals, geophysical activity, and to human impacts on the environment. During the same time frame (1.64 million years ago to the present), islands probably suffered extinctions from each of these causes as well. During the lowered sea levels of glacial maxima, the shallow marine shelves that surround some islands were exposed as emergent land, sometimes forming land bridges to continents or to other islands. Conversely, islands with low physical relief have been partly (sometimes entirely) inundated by the sea during marine transgressions. In theory, extinctions can result from competition after faunal exchange in the former case, or from reduced carrying capacity in the latter.

Knowledge of the historical geology of islands is receiving the benefit of renewed interest in catastrophism, leading to intriguing research, for example on the immense submarine (and partly subaerial) debris avalanches that affect Hawaiian shield volcanoes. Enormous avalanches can cause a sudden reduction in island area, and give rise to waves that have reached up to 365 m elevation on nearby islands, stripping away top soil below this elevation (Moore et al. 1989). Sudden devastation to islands can result from volcanic eruptions as well. Such catastrophic geophysical events have the potential to cause significant disturbance, and even to lead to extinctions on islands.

Delayed vegetation change following the Holocene climatic reversal may explain the extinction of dwarf mammoths 4000 years ago on Wrangel Island in the Arctic Ocean (Vartanyan et al. 1993). Change from drier Pleistocene to wetter Holocene climate has been advanced to explain extinctions in the West Indies (Pregill and Olson 1981), although chronological data are needed to support this. In Madagascar, paleoecological evidence of climatic desiccation during the late Holocene, coupled with evidence for several types of human impacts, has given rise to the hypothesis that synergistic interactions among climatic and anthropogenic factors contributed to the extinctions (e.g., MacPhee 1986; Burney 1993a). This explanation has been suggested for some late Quaternary extinctions outside Madagascar as well (Burney 1993b).

The most widely held hypotheses to explain Holocene island extinctions include a strong component of anthropogenic change. When the fossils have been radiometrically dated, it is usually apparent that most extinctions postdate the arrival of humans on the island (e.g., Steadman 1989; Steadman and Olson 1985; Steadman et al. 1984). Overharvesting and habitat destruction by humans, and collateral impacts resulting from their introduction of foreign competitors, predators, and even pathogens, are thought to have played a role. The human-caused extinction of island endemics is an ongoing phenomenon, and a matter of historical record in many instances (Greenway 1958). For prehistoric extinctions, the most thorough demonstrations of the human role have come from multidisciplinary studies that associate vertebrate fossils with evidence drawn from sedimentology, palynology, archeology and ethnography (e.g., Anderson 1989; Kirch et al. 1992).

For island ecosystem studies, it is important to understand that prehistoric anthropogenic change has intervened between modern patterns and the long-term processes that we infer from them. To what extent can modern patterns of diversity on islands be attributed to natural factors, to what extent to anthropogenic change? What exactly caused the extinctions, and how was ecosystem function affected? I review data relevant to some aspects of these questions, and where further research is needed, discuss the types of paleontological data that can be brought to bear.

8.2 Diversity

Holocene extinctions of vertebrates have been detected on most islands where fossil surveys were carried out. The proportion of native vertebrates lost from individual islands varies, but in the most extreme examples, some islands lost all resident, terrestrial species (e.g., Easter Island and Kahoolawe). New Zealand lost about 30 species of birds prehistorically (Millener 1991). The Hawaiian Islands lost 35 to perhaps 57 (James and Olson 1991; Olson and James 1991), an estimate that is continually being revised upward as more new species are discovered (e.g., Giffin 1993). Madagascar lost at least 30 species of mammals, birds, and tortoises (Dewar 1984; Godfrey et al. 1990; Goodman and Ravoavy 1993; and James pers. observe). In recent centuries, even the relatively pristine Galapagos Islands have lost at least 29 to 34 island populations of vertebrates (Steadman et al. 1991).

The Hawaiian Islands can serve to illustrate how prehistoric extinctions obscured our perception of natural faunal development on islands. When the extinct birds are added to the avifauna, it becomes apparent that many more avian species successfully colonized the archipelago and gave rise to endemic lineages than would be recognized without the fossil evidence (Fig. 8.1). The lineages that became extinct in prehistoric times fall into two distinct categories. They were either derived from waterbirds (ibises, rails, geese, ducks), most of which had shifted to terrestrial habitats after colonizing the islands, or they were

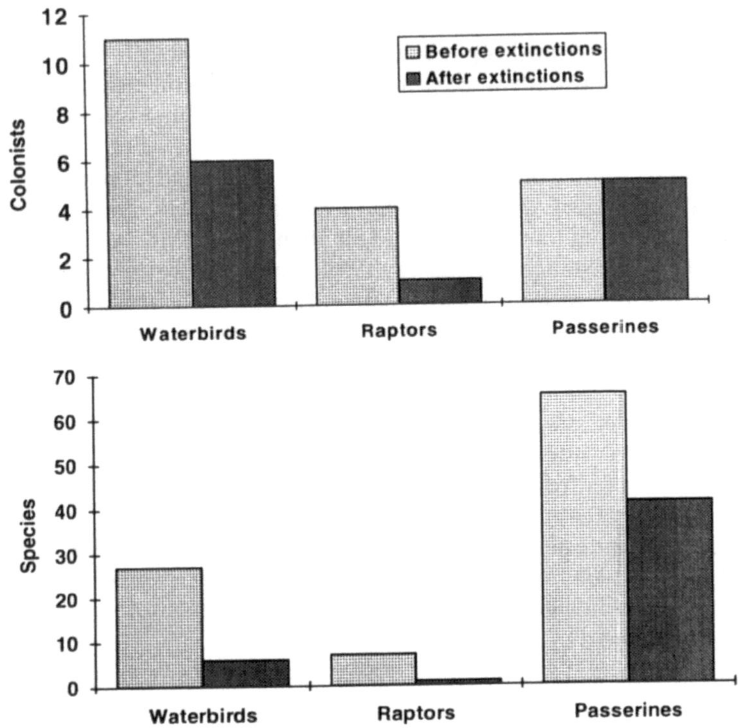

Fig. 8.1. The native avifauna of the main Hawaiian Islands, before and after prehistoric extinctions. *Above* The number of colonizing species necessary to account for the resident species of birds; *below* the number of resident species of birds. Data on the number of colonists are from James (1991). Note that all successful colonists were either waterbirds, raptors, or passerines. Many of the waterbirds shifted into terrestrial habitats after colonizing the islands

raptors (eagles, hawks, owls). No passerine lineages became entirely extinct, and passerines suffered proportionately less extinction at the species level as well (Fig. 8.1, below). In this way, extinction masked the success of waterbirds and raptors in colonizing the remote Hawaiian Islands and diversifying there, while it artificially exaggerated the importance of passerines in the fauna.

The impact that extinction may have had on individual island ecosystems is suggested by Figure 8.2, which contrasts the number of resident, native species recorded from specific Hawaiian islands, before and after prehistoric extinctions. Fossil collecting has been the most thorough on Maui and Oahu. On these islands, the high proportion of extinct species suggests that a prehistoric faunal collapse occurred, affecting not only the larger species that are conventionally thought to be vulnerable, but also many smaller species that presumably would have been relatively abundant and less extinction-prone.

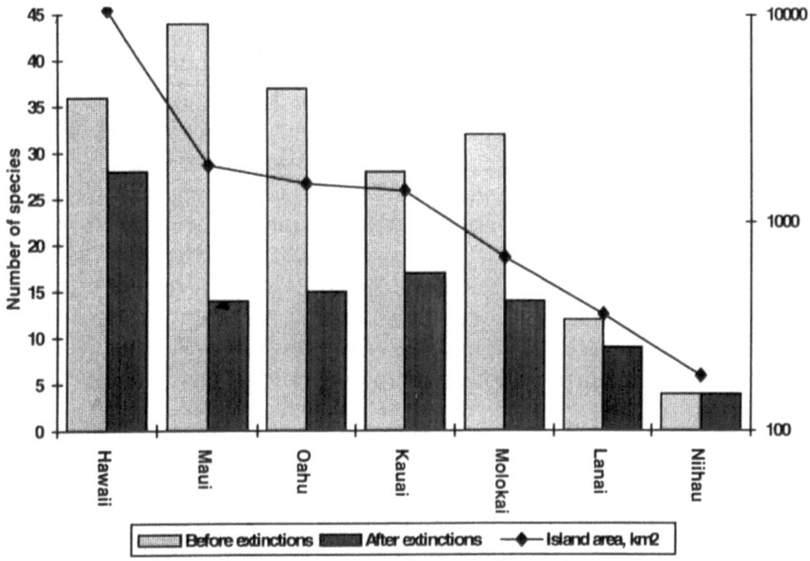

Fig. 8.2. The number of species in the resident avifauna of individual Hawaiian islands, before and after prehistoric extinctions. The scale on the *right* is for island area

Table 8.1. Comparison of cave faunas from Maui and northwest Madagascar

	Madgascar	Maui
Island area, km^2	5.87×10^5	1.18×10^3
Human arrival, thousands of years before present	$\leqslant 2$	$\leqslant 2$
Local climate near caves	Tropical dry/mesic, seasonal rainfall	Tropical dry/mesic, seasonal rainfall
Cave system	Grottes d'Anjohibe and Anjohikely (S15° 33', E46° 53)	Puu Naio Cave (N20° 37', W156° 24')
Chronological range of fossil assemblage	7790 B.P. to modern	8820 B.P. to modern
Species of higher vertebrates		
n identified	±65	35
n extinct	7–10	24–26
%extinct	11–15	69–74

In other island ecosystems the extinction event may have had a substantially different profile. A comparison of cave faunas from northwest Madagascar and Maui illustrates this point (Table 8.1). Radiocarbon chronologies establish that bones were accumulating in the two cave systems (Puu Naio on Maui, and Anjohibe and Anjohikely on Madagascar) during most of the Holocene. The taphonomic origin of the collections is from natural traps and predator accumulations, supplemented with samples from bat roosts and archeological sites in the Madagascar caves. Thousands of bones were collected and identified from each cave system (James et al. 1987; James and Olson 1991; Burney et al. 1994). The revealing comparison is between the number and proportion of extinct species in the two localities. In northwest Madagascar, the evidence suggests that extinctions affected a much smaller proportion of the fauna and were more skewed toward larger species, compared to what occurred on Maui. If the model of faunal collapse with cascading extinctions is applicable to both extinction events, then the cascade appears to have proceeded much farther on Maui. Alternately, the difference may be explained to some extent by sampling error. Further field work may reveal a large number of undocumented extinctions of smaller animals on Madagascar.

Little is known about prehistoric extinctions of animals other than vertebrates. Insect extinctions probably occurred, but with few exceptions, there is no relevant record. Remains of extinct land crabs are frequently found in dunes and caves in the Hawaiian Islands. These may be the same species that was collected alive on Oahu in 1864 (Edmonson 1962), a record that has generally been overlooked because land crabs were never again taken alive in the islands. Land snails likewise suffered extinctions and local extirpations, and left behind an abundant record in caves and sediments, worthy of further study (e.g., Christensen and Kirch 1986).

The island record of fossil and subfossil plants generally has not produced evidence of anthropogenic extinction on the same scale as for vertebrates, but there are notable exceptions. For example, no native trees live on Easter Island, but lake cores contain pollen types representing about seven arborescent taxa (Flenley et al. 1991). Their extinction is attributed to complete deforestation of the island by humans. Ilha da Trindade in the South Atlantic is now virtually barren of trees, but dried stumps and branches found around the island vouchsafe the formerly wide distribution of a tree that was mentioned in early travel accounts (Eyde and Olson 1983). In the Hawaiian Islands, most of the Holocene plant microfossils identified by Selling (1946, 1947, 1948) in mid- to high-elevation bog cores belonged to extant taxa, although the spores of one fern were described as an extinct species (Selling 1946). The pollen types identified in late-Holocene sediment cores from the lowlands of Oahu were also attributed to extant taxa (Athens et al. 1992). However, the lowland cores provided evidence that there were major changes in plant communities during prehistoric times. These changes included the island-wide extinction of a leguminous shrub that had been fairly common on Oahu in the late Holocene. As one of the few nitrogen-fixing plants in the native Hawaiian flora, this legume may once have played an ecologically important role, but it is now too rare to do so. To summarize the botanical evidence, it is clear that prehistoric anthropogenic changes in island vegetation may have been extensive and have had important implications for ecosystem function. Yet current evidence suggests that plants were more resistant to extinction than animals (at least vertebrates) in the same ecosystems.

8.3 Prehuman Extinctions

In order to understand how island ecosystems respond to disturbance, be it natural or anthropogenic, we first need to know how these ecosystems behave in the absence of disturbance. When the evidence of disturbance is mainly a record of extinctions, we need to know the frequency of background extinction for comparison. In ecology, background extinctions are theoretically linked to species immigrations, which are often combined to calculate the overall rate of species turnover. Based on ecologists' estimates of species turnover in island avifaunas, summarized by Schoener (1983), we might expect 500 to 14000 turnover events on a given island in the course of the Holocene.

For a number of reasons, it is likely that oceanic island communities are far more stable than these rates imply. Not all bird censuses yield similarly high estimates of turnover (Mayer and Chipley 1992). Arguing from modern patterns of diversity, we might predict low natural turnover rates on many oceanic islands. Turnover is inferred to be low when a high proportion of the species on an island are endemic, since endemism cannot develop or be sustained unless populations on the island are fairly long-lived (Diamond and Jones 1982). However, we have already discussed the prehistoric changes that make us suspicious of inference from modern patterns.

Estimating the turnover rate from ecological data has always been controversial. Early attempts were faulted because they relied on incomplete census data (Lynch and Johnson 1974), and attached undue importance to "ecologically trivial" events, such as the failure of small founder populations that had never played a significant role in ecosystem dynamics (Williamson 1989). In some studies of bird populations, most of the observed turnover was due to occasional breeders that were recorded in some years but not others (Diamond and Jones 1982). Considered over a longer stretch of time, these species are effectively part of the island's avian community. It may be inappropriate to equate their movements with turnover of long-established resident populations. In addition to these drawbacks, widespread anthropogenic disturbance makes it difficult to find island ecosystems where turnover can be studied under natural conditions. All of the above-mentioned factors will tend to inflate turnover rates estimated from ecological data, when compared with paleontological data.

Consider now the characteristics of paleontological data with respect to estimating turnover. In the light of time-averaging of fossil assemblages, census interval poses an even more serious problem than it does for ecological data (Diamond and May 1977). Time-averaging also means that paleontological records are unlikely to detect turnover due to occasional breeders. Small founder populations that fail to take hold are probably less likely to be observed by the paleontologist than by the field ecologist. Distinguishing breeding from non-breeding birds in fossil samples can be an additional problem. In these ways, paleontological and ecological estimates of turnover rate will not be directly comparable. However, if we wish to know to what extent extinctions and immigrations have shaped the modern ornithogeography of islands, then paleontological data have some advantages. Only fossils offer the time-depth necessary to gauge the importance of turnover for long-established, resident populations. If such populations experience a constant rate of background extinction, this should be revealed by paleochronological data that record the distribution of extinctions through time.

Most available data show a cluster of extinctions in the human era, and very few before then. For example, in the Galápagos, of the 29 to 34 extinctions of vertebrate species or island populations that are on record for the Holocene, all but 3 are known to have occurred in the 5 centuries since humans began visiting the islands (Steadman et al. 1991).

Taking a step back in time, on the west coast of the South Island, New Zealand, avian species composition in cave faunas differs significantly between the last glacial maximum of the Pleistocene (25 000 to 10 000 years ago) and the Holocene (Worthy and Holdaway 1993). The authors interpret these differences in terms of the Pleistocene-Holocene climatic reversal. Evidently, climatic reversal caused major regional shifts in bird distributions within New Zealand, but as far as we know, no outright extinctions occurred. All of the late Pleistocene fossils of terrestrial birds known so far from New Zealand are attributed to species that survived until humans arrived, about 1000 years ago (Cooper and Millener 1993). The New Zealand avifauna suffered far greater stress during the human era, when half of the terrestrial species went extinct.

Taking another step back in time, on the island of `Eua, Tonga, 21 species of birds were identified in a fossil collection dating to between 60 000 and 80 000 years ago (Steadman 1993). The fate of these species was traced by checking for their presence in archeological deposits on `Eua, which data to < 3000 years ago. All but 5 of the 21 species were recorded in archeological contexts, indicating that at least three-fourths of the Pleistocene fauna survived until the human era. Eleven of these survivors (half of the Pleistocene fauna) disappeared from `Eua during the human era. The Tongan data are consistent with the pattern of higher extinction rates in the human era, but they also suggest more prehuman faunal change than was observed in New Zealand. The author of the study notes that some of the apparent prehuman extinctions may be artifacts of sampling error.

A deeper Pleistocene record is available from the Hawaiian island of Oahu, where 17 species of land birds were recorded in a preliminary study of Pleistocene fossils from Ulupau Head, dating to > 120 000 years ago (James 1987). The fate of these species can be traced up through the Holocene to the present. Between > 120 000 years ago and the present, global climate passed through a complete cycle of glaciation and deglaciation, a potential cause of faunal change. However, in this case very little faunal change was observed in prehuman times. All but perhaps one or two (12% or fewer) of the species from Ulupau Head survived through the times of most pronounced climate change. These species were still extant in the mid- to late Holocene, when the Barber's Point fossil assemblage accumulated on Oahu. Tellingly, 13 (76%) of the Pleistocene survivors have met with extinction during the past few thousand years, when human impacts may have been important.

In summary, although we cannot hope to calculate annual turnover rates for comparison with ecological data, the paleontological evidence argues for long-term stability of vertebrate communities on oceanic islands. There is more evidence of prehuman faunal change in Tonga than in Hawaii or New Zealand. In New Zealand and Hawaii, the fossil evidence would not support species turnover as an important determinant of modern bird distributions, at least for native species. Perhaps more surprising is the paucity of evidence for climate-driven extinctions in the late Pleistocene, whether in tropical Hawaii or in temperate to subantarctic New Zealand. There are at least two possible explanations for this. One is simply that some island ecosystems resist climate-driven extinctions better than some continental ecosystems. Another is that late Pleistocene climate change actually caused relatively few extinctions, either on continents or on islands. It is still debated that late Pleistocene extinctions on continents were caused mainly by the spread of paleolithic peoples, rather than by climatic reversal (Martin 1984).

8.4 Prehuman Biological Invasions

Humans can cause disturbance on oceanic islands through their own activities, but they also frequently assist other alien species to invade, sometimes with very deleterious results. However, we know that island ecosystems have been invaded by alien species many times in the past, and presumably suffered adverse consequences even before the advent of humans. To put human-assisted invasions in perspective, it would be useful to know how island ecosystems have responded to invasions when humans were absent. This is another aspect of island ecosystem dynamics that paleontology may help to illuminate.

For example, several lines of evidence suggest that the subfossil sea eagle of the Hawaiian Islands is a relatively recent, but prehuman invader. First, the Hawaiian eagle does not show morphological adaptations to the island environment. Osteologically, it cannot be distinguished from either the palearctic White-Tailed Sea Eagle (*Haliaeetus albicilla*), or from the nearctic Bald Eagle (*Haliaeetus leucocephalus*; see Olson and James 1991). Second, chronological data are consistent with recent colonization. Collagen from eagle bones found in Puu Makua Cave on Maui was dated as being over 3000 years old (^{14}C sample 88NM-22, 3000 ± 60 B. P.; 2 sigma calibrated range, 3389–3689 B. P.). This date establishes the presence of the eagle in the islands at least 1300 years before human arrival. More tentatively, the eagle may not have been present as long ago as > 120 000 years, as it has not been collected from the Pleistocene lake beds at Ulupau Head. We know that most of the elements of the Hawaiian vertebrate fauna (such as large flightless anseriforms, small ducks, coots, hawks, owls, crows, and flightless rails) were already present > 120 000 years ago (James 1987). The subsequent arrival of an eagle would have introduced a new trophic level to Hawaiian ecosystems, where flightless birds may have evolved in the absence of large predators. If adjustments occurred in response to predation by the eagle, there is no evidence that they included rapid extinctions. Radiocarbon chronologies indicate that the extinctions of flightless birds took place centuries after humans arrived, and at least 2000 years after the arrival of the eagle (Stafford and James, unpubl.).

It is possible to cite contrary examples in which the human-assisted invasion of an island by a predator leads to rapid extinctions, as when the brown tree snake was transported to Guam (Savidge 1987). Further development of the fossil record associated with prehuman invasions of oceanic islands would lend historical perspective to the many studies of biological invasions that are underway at this time.

8.5 Anthropogenic Deletions

The ecological effects of biological invasions have received more research attention than the effects of deletions, perhaps not surprisingly in light of the difficulty of studying something that is no longer there. Nevertheless, deletions should not be ignored, since the fossil record repeatedly points to their importance in shaping modern communities. Consider the rainforest of the Kipahulu Valley on Maui, one of the few seemingly pristine forests remaining in the Hawaiian Islands. Seven species of birds that no longer occur in the valley constitute the faunal list of Holocene fossils from lava tube caves here (Medeiros et al. 1988; James, pers. observ). Only one of these species, the Hawaiian dark-numped petrel (*Pterodroma phaeopygia*), still survives anywhere on the island. The forest has also lost an herbivorous flightless anatid (*Ptaiochen pau*), a goose related to the Hawaiian Goose but with reduced flight capability (*Branta hylobadistes*), a flightless rail (*Porzana* sp.), a flightless ibis (*Apteribis* sp.), a thrush (*Myadestes* sp.), and a finch (*Orthiospiza howarthi*). The fossil sample is small and the list is undoubtedly incomplete, yet even with this limited information, it is evident that the valley has undergone considerable Holocene change. Although the exact dietary habits of the missing species are unknown, we can surmise that the forest has been released from pressure from herbivores, frugivores, probably granivores, and from two species that were probably omnivorous feeders among forest floor detritus. What specifically caused these species' disappearance from the valley is not known, but human impacts of some sort are a strong possibility.

On many islands, prehistoric deletions may still be affecting forest ecosystems in unseen ways. Relevant observational data are hard to come by, leaving this a subject for current speculation and future research. Some possible effects of prehistoric deletions are the loss of certain ecological roles (e.g., with the extinction of land crabs), the breakdown of mutualisms (e.g., between rare plants and their extinct pollinators), habitat change due to loss of keystone species (e.g., the New Zealand moas, whose extinction may have inhibited the regeneration of native conifers; Cooper et al. 1993), and limitations to plant dispersal (e.g., due to the widespread extinction of avian frugivores in Polynesia; Steadman 1991). A factor worth investigating is the role that breeding seabirds may have played in transferring marine nutrients to terrestrial ecosystems. Breeding seabirds were extirpated in some regions and greatly reduced in numbers in others by prehistoric overharvesting.

8.6 Vertebrate Feeding Guilds

Compared to the possible effects mentioned above, prehistoric changes to vertebrate feeding guilds are fairly amenable to analysis. Patterns can be discerned, for instance that extinctions left a false impression of relatively predator-free communities on many islands. From giant owls in Cuba, to giant eagles in Crete and New Zealand, to foxes in Sardinia and Corsica, predators at the highest trophic level seem to have been particularly vulnerable to prehistoric extinction. Large, endemic herbivores were vulnerable as well. Moas, swans, and geese are gone from New Zealand; elephant birds, giant tortoises, pygmy hippos, and giant lemurs from Madagascar; ground sloths and monkeys from the Greater Antilles. Many, but not all, of these extinctions have been shown to postdate human arrival.

The Hawaiian Islands can serve once again to illustrate how prehistoric extinctions altered the structure of vertebrate feeding guilds. Based on a very rough division of the resident avifauna into guilds, it is clear that extinction affected some groups more than others (Fig. 8.3). All but one species of native predator became extinct prehistorically. The one survivor (the Hawaiian Hawk, *Buteo solitarius*) became restricted in distribution to the largest island, so that most regions of the islands lost all native predators. To some extent, native raptors were replaced by owls that arrived after human settlement, but the new arrivals tend to feed on rodents, while their extinct counterparts specialized in eating the native birds.

Similarly, all terrestrial herbivores became extinct except the Hawaiian goose (*Branta sandvicensis*), which survived only on the largest island. The extinct herbivores include the large flightless anseriforms called moa-nalos, which are thought to have been browsers on understory foliage (Olson and James 1991). Their universal extinction may still be affecting understory vegetation in Hawaiian forests. Terrestrial omnivores (flightless rails and ibises) had been an important component of Hawaiian ecosystems, but they too all but disappeared prehistorically. The absence of avian terrestrial omnivores, coupled with the extinction of Hawaiian land crabs, may have continuing implications for nutrient cycling on the forest floor.

Most of the species of vertebrates that survived the Hawaiian extinction event are perching birds (order Passeriformes). Of 67 passerine species in the Holocene fauna of the main islands, 25 (37%) became extinct prehistorically, and another 15 (22%) have become extinct since European contact. These figures are approximate, because some postcontact extinctions probably went undetected, and undoubtedly some prehistoric extinctions have not yet been discovered.

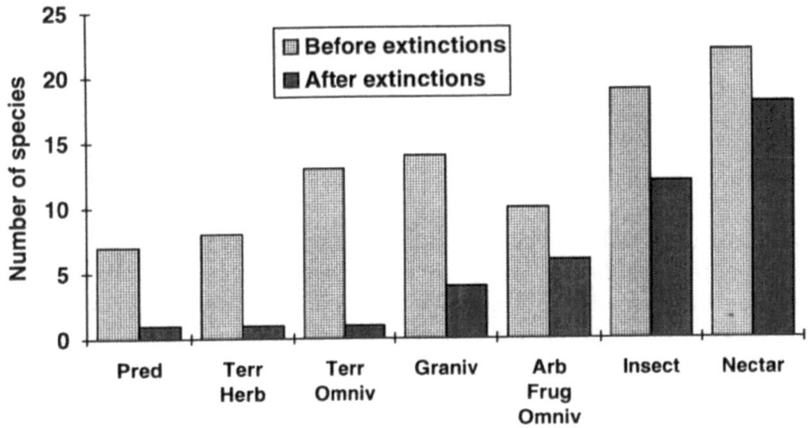

Fig. 8.3. The effect of extinction on avian feeding guilds in the Hawaiian Islands. Guilds are as follows: predators (*Haliaeetus, Buteo, Circus, Grallistrix*); terrestrial herbivores (*Chelychelynechen, Thambetochen, Ptaiochen, Branta, Geochen*, very large Hawaii goose); terrestrial omnivores (*Apteribis, Porzana*); granivores (*Telespiza, Loxioides, Chloridops, Rhodacanthis, Xestospiza*, and questionably *Orthiospiza*); arboreal frugivores/omnivores (*Corvus, Myadestes, Psittirostra*); insectivores (*Chasiempis, Pseudonestor, Hemignathus, Oreomystis, Paroreomyza, Vangulifer, Aidemedia*); and nectarivores (*Moho, Chaetoptila, Loxops, Vestiaria, Drepanis, Himatione, Palmeria, Ciridops*). Wetland species and non-native species are excluded from the analysis. Snail-eaters (*Melamprosops* and questionably *Dysmorodrepanis*) are excluded from the figure

Not enough is known about the diets of the extinct species to permit a precise division of the passerines into feeding guilds at this time. Instead, I have divided them into four very general categories, an approach that will reveal only the broadest patterns (Fig. 8.3). Examined in this way, prehistoric extinctions among insectivores (37%) and arboreal frugivores/ominivores (40%)are proportional to extinctions among passerines generally.

The same is not true of granivores or nectarivores. Granivores have suffered grievously from extinction, beginning in the prehistoric phase of extinctions, and continuing historically until they were obliterated from all but one area of the main islands, high on the slopes of Mauna Kea. Meanwhile, nectarivores enjoyed an opposite fate. They began as the most speciose guild and remained so. Their extinctions were especially light during the prehistoric phase, when most of the granivores were disappearing. Indeed, the changes that Hawaiian forest communities have undergone seem to favor certain nectarivores (i.e., *Himatione sanguinea, Loxops virens, Vestiaria coccinea*), which at the present time are the most abundant native birds in the forests (Scott et al. 1986). With predators disappearing, habitats shifting, and species richness plummeting during the extinction event, it is plausible to suppose that some species, even some guilds, might gain an advantage, and perhaps end up being more widespread and abundant than they had been under prehuman conditions.

8.7 Future Directions

More than any other factor, the Holocene fossil record of islands reflects the repercussions of one significant addition to the fauna, that of *Homo sapiens*. As humans became established on island after island, an extinction event ensued. In effect, the experiment that can tell us how ecosystems respond to anthropogenic change, and particularly to reduced species richness, has already been set in motion. The experiment has been replicated numerous times on different islands, under varying conditions of climate, geography, ecosystem structure, human economy, and cultural evolution. The challenge is to exploit this source of information more effectively, to better understand what caused the extinctions, and how ecosystems were affected.

Reconstructing the complex history of prehistoric environmental degradation and identifying its precise causes is admittedly a difficult task. The use of comparative methods to search for common patterns in the extinction events that occurred on different islands is one promising approach to the problem. For this approach to be successful, we need to develop comparable data sets from various islands, dealing with such factors as vegetation change, human population and adaptation, and the ecological characteristics of extinct vs. surviving species. Excellent chronological data are needed, so that events that were separated in time will not be falsely correlated with each other. The comparative data on extinctions in northwest Madagascar and Maui, discussed in this chapter, is one small example of the types of comparisons that could prove to be informative.

The differential effects of extinction on vertebrate feeding guilds suggest that prehistoric changes may still be affecting island communities. When extinctions have occurred but the forest itself has endured, as is the case in the Kipahulu Valley on Maui, there may be continuing effects on ecosystem dynamics that are subtle and difficult to study. What happens to a forest when the native vertebrate predators, terrestrial herbivores and omnivores, and the granivores are removed from it? Increased collaboration between ecologists and paleontologists would promote a better understanding of natural ecosystem processes in the native habitats that still persist on islands.

A final point for emphasis is the potential of the fossil record to illuminate island ecosystem dynamics in prehuman times. Our perception of island ecosystems as fragile, and of island endemics as extinction-prone, is colored by the accelerated pace of change during the human era. In prehuman times, vertebrate communities on some islands underwent very little reorganization for tens of thousands of years, even through periods of pronounced global climate change. Stasis, however, is not the entire story. Before the advent of humans, island ecosystems were exposed to stress, for example from naturally occurring biological invasions, from reconfiguration of the landscape due to sea level change, from catastrophic geophysical events, and certainly from changes in climate. By developing the fossil record associated with such events, we could improve our understanding of the effects of natural vs. anthropogenic ecosystem disturbance.

Acknowledgments. I am grateful to Peter Vitousek and Hall Cushman of Stanford University and to Pericles Moillis of the Bahamas National Trust for organizing a very rewarding workshop. The field research that led to this summation was supported by the Wetmore Funds and the Scholarly Studies Program of the Smithsonian Institution, the Department of Defense Legacy Program, National Geographic Society grant 4493–91 to H. Wright and D. Burney, and NSF BSR-9025020 to D. Burney. David Burney, Lloyd Loope, Christopher Perrins, Storrs Olson, David Steadman, and Graham Wragg read a draft of the manuscript and made many useful suggestions.

References

Anderson A (1989) Prodigious birds: moas and moa-hunting in prehistoric New Zealand. Cambridge University Press, Cambridge

Athens JS, Ward J, Wickler S (1992) Late Holocene lowland vegetation, O'ahu, Hawai'i NZJ Archaeol 14: 9–34

Atkinson IAE, Millener PR (1991) An ornithological glimpse into New Zealand's pre-human past. Acta XX Congr Int Ornithol 1: 127–192

Berger AJ (1972) Hawaiian birdlife. University of Hawaii Press, Honolulu, xii + 270 pp

Burney DA (1993a) Late Holocene environmental changes in arid southwestern Madagascar. Quat Res 40: 98–106

Burney DA (1993b) Recent animal extinctions:recipes for disaster. Am Sci 81: 530–541

Burney DA, James HF, Grady FV, Rafamantanantsoa J-G, Ramilisonina, Wright HT, Cowart JB (1994) Environmental change, extinction, and human activity: evidence from caves in NW Madagascar. Natl Geogr Res Explor (in press)

Christensen CC, Kirch PV (1986) Nonmarine mollusks and ecological change at Barbers Point, O'ahu, Hawai'i. Occas Pap Bernice P Bishop Mus 26: 52–80

Cooper RA, Millener PR (1993) The New Zealand biota: historical background and new research. TREE 8(12): 429–433

Cooper AJ, Atkinson IAE, Lee WG, Worthy TH (1993) Evolution of the Moa and their effect on the New Zealand flora. TREE 8(12): 433–437

Dewar RE (1984) Extinctions in Madagascar: the loss of the subfossil fauna. In: Martin PS, Klein RG (eds) Quaternary extinctions: a prehistoric revolution. University of Arizona Press, Tucson, pp 574–593

Diamond JM, Jones HL (1982) Species turnover in island bird communities. Acta XVII Congr Int Ornithol 2: 777–782

Diamond JM, May RM (1977) Species turnover rates on islands: dependence on census interval. Science 197: 266–270

Edmonson CH (1962) Hawaiian Crustacea: Goneplacidae, Pinnotheridae, Cymopoliidae, Ocypididae, and Gacardinidae. Occas Pap Bernice P Bishop Mus 23(1): 1–27

Eyde RH, Olson SL (1983) The dead trees of Ilha da Trindade. Bartonia 49: 32–51

Flenley JR, King ASM, Jackson J, Chew C (1991) The Late Quaternary vegetational and climatic history of Easter Island. J Quat Sci 6(2): 85–115

Giffin J (1993) New species of fossil birds found at Pu'u Wa'awa'a. `Elepaio 53(1): 1–3

Godfrey LR, Simons EL, Chatrath PJ, Rakotosamimanana B (1990) A new fossil lemur, *Babakotia* (Primates), from northern Madagascar. CR Acad Sci ser II Mec Phys Chim Sci, Univers Sci Tone 310(1): 81–88

Goodman SM, Ravoavy F (1993) Identification of bird subfossils from cave surface deposits at Anjohibe, Madagascar, with a description of a new giant *Coua* (Cuculidae:Couinae). Proc Biol Soc Wash 106(1): 24–33

Greenway JC (1958) Extinct and vanishing birds of the world. Spec Publ 13, Am Comm Int Wild Life Protect New York

James HF (1987) A late Pleistocene avifauna from the island of Oahu, Hawaiian Islands. Doc Lab Geol Fac Sci Lyon 99: 221–230
James HF (1991) The contribution of fossils to knowledge of Hawaiian birds. Acta XX Congr Int Ornithol 1: 420–424
James HF, Olson SL (1991) Descriptions of thirty-two new species of birds from the Hawaiian Islands: Part II. Passeriformes. Ornithol Monogr 46: 1–88
James HF, Stafford TW Jr, Steadman DW, Olson SL, Martin PS, McCoy P (1987) Radiocarbon dates on bones of extinct birds from Hawaii. Proc Natl Acad Sci USA 4: 2350–2354
Kirch PV, Flenley JR, Steadman DW, Lamont F, Dawon S (1992) Ancient environmental degradation: prehistoric human impacts on an island ecosystem: Mangaia, Central Polynesia. Natl Geogr Res Expl 8(2): 166–179
Lynch JF, Johnson NK (1974) Turnover and equilibria in insular avifaunas, with special reference to the California Channel Islands. Condor 76(4): 370–384
MacPhee RDE (1986) Environment, extinction, and holocene vertebrate localities in southern Madagascar. Natl Geogr Res Expl 2(4): 441–455
Martin PS (1984) Prehistoric overkill: the global model. In: Martin PS, Klein RG (eds) Quaternary extinctions: a prehistoric revolution. University of Arizona Press, Tucson, pp 354–403
Mayer GC, Chipley RM (1992) Turnover in the avifauna of Guana Island, British Virgin Islands. J Anim Ecol 61: 561–566
Medeiros AC, Loope LL, James HF (1989) Caves, bird bones and beetles: new discoveries in rain forests of Haleakala. Park Sci 9(2): 20–21
Milberg P, Tyrberg T (1993) Naive birds and noble savages – a review of man-caused prehistoric extinctions of island birds. Ecography 16: 229–250
Millener PR (1991) The Quaternary avifauna of New Zealand. In: Vickers-Rich PV, Monaghan JM, Baird RF, Rich TH (eds) Vertebrate paleontology of Australasia. Monash Univ Publ Comm, Melbourne, pp 1317–1344
Moore JG, Clague DA, Holcomb RT, Lipman PW, Normark WR, and Torresan ME (1989) Prodigious submarine landslides on the Hawaiian Ridge. J Geophys Res 94 (B12): 17465–17484
Olson SL, James HF (1991) Descriptions of thirty-two new species of birds from the Hawaiian Islands: Part I. Non-Passeriformes. Ornithol Monogr 45: 1–88
Pregill GK, Olson SL (1981) Zoogeography of West Indian vertebrates in relation to Pleistocene climatic cycles. Annu Rev Ecol Syst 12: 75–98
Savidge JA (1987) Extinction of an island forest avifauna by an introduced snake. Ecology 68: 660–668
Schoener TW (1983) Rate of species turnover declines from lower to higher organisms: a review of data. Oikos 41: 372–377
Scott JM, Mountainspring S, Ramsey FL, Kepler CB (1986) Forest bird communities of the Hawaiian Islands: their dynamics, ecology, and conservation. Stud Avian Biol 9: 1–431
Selling OH (1946) Studies in Hawaiian pollen statistics, Part I: the spores of Hawaiian Pteridophytes. Spec Publ Bishop Mus 37: 1–87
Selling OH (1947) Studies in Hawaiian pollen statistics, part II: the pollens of the Hawaiian phanerogams. Spec Publ Bishop Mus 38: 1–430
Selling OH (1948) Studies in Hawaiian pollen statistics, part III: on the late Quaternary history of the Hawaiian vegetation. Spec Publ Bishop Mus 39: 1–154
Steadman DW (1989) Extinction of birds in eastern Polynesia: a review of the record, and comparisons with other Pacific island groups. J Arch Sci 16: 177–205
Steadman DW (1991) Ecological impact of the human depletion of frugivorous birds in Polynesia. Acta XX Congr Int Ornithol 1: 424 (Abs)
Steadman DW (1993) Biogeography of Tongan birds before and after human impact. Proc Natl Acad Sci USA 90: 818–822
Steadman DW, Olson SL (1985) Bird remains from an archaeological site on Henderson Island, South Pacific: man-caused extinctions on an "uninhabited" island. Proc Natl Acad Sci USA 82: 6191–6195

Steadman DW, Pregill GK, Olson SL (1984) Fossil vertebrates from Antigua, Lesser Antilles: evidence for late Holocene human-caused extinctions in the West Indies. Proc Natl Acad Sci USA 81: 4448–4451
Steadman DW, Stafford TW Jr, Donahue DJ, Jull AJT (1991) Chronology of Holocene vertebrate extinction in the Galapagos Islands. Quat Res 36: 126–133
Vartanyan SL, Garutt VE, Sher AV (1993) Holocene dwarf mammoths from Wrangel Island in the Siberian Arctic. Nature 362: 337–340
Williamson M (1989) The MacArthur and Wilson theory today: true but trivial. J Biogeogr 16: 3–4
Worthy TH, Holdaway RN (1993) Quaternary fossil faunas from caves in the Punakaiki area, West Coast, South Island, New Zealand. J R Soc NZ 23(3): 147–254

Note added in proof. The manuscript cited above as Burney et al. 1994 (in press) was accepted for publication, but the journal in question recently ceased publication and all manuscripts were returned to the authors. Another publication outlet is being sought.

9 Biological Invasions as Agents of Change on Islands Versus Mainlands

C. M. D'ANTONIO[1] and T. L. DUDLEY[2]

9.1 Introduction

Over the past decade, the Scientific Committee on Problems of the Environment (SCOPE) has devoted considerable attention to the ecology of invasive organisms (e.g., Brockie et al. 1988; Loope et al. 1988; MacDonald and Frame 1988; MacDonald et al. 1988; Usher 1988; Drake et al. 1989). Conclusions emerging from the workshops that fostered these publications include: (1) perhaps most importantly – biological invasions are now recognized as a global phenomenon that can affect both structure and function in ecosystems; (2) even nature reserves, typically pieces of land set aside because they represent intact native species-dominated ecosystems, are subject to serious biological invasions; (3) island ecosystems typically have a higher representation of alien species in their flora and fauna than do mainland systems; and (4) the severity of invasions on islands increases with isolation of the island. The latter two conclusions are consistent with what has become dogma in discussions of invading species – that island communities are more invasible than mainland ones. Investigators have often expressed this viewpoint with statements such as "because it evolved in isolation, this native fauna [of islands] is highly susceptible to the introduction of alien species" (Richardson 1992). This statement does not distinguish between the possibilities that alien species might establish more readily on islands versus that alien species might have more of an effect (population, community level, etc.) on islands than they do on continents. In other words, processes allowing or promoting invasion are not distinguished unambiguously from the impacts of invaders once they have become established.

In this chapter, we would like to separate the concept of community "invasibility" from that of *the effects* invaders are having on island communities. We will first reexamine the idea that island habitats are inherently more easily invaded than are continental habitats and argue that, with the exception of some taxa, this is to a large extent a poorly supported contention. We will argue that regardless of the ultimate causes of invasion, island species and island ecosystems are often more susceptible to the effects of invaders. This has probably contributed to the impression that islands are more "invasible". We will speculate as to why island

[1] Department of Integrative Biology, University of California, Berkeley, CA 94720, USA
[2] Pacific Institute for Studies in Development, Environment and Security, 1204 Preservation Park Way, Oakland, CA 94612, USA

Ecological Studies, Vol. 115
Vitousek et al. (eds) Islands

species and ecosystems appear to suffer greater impacts from introduced organisms than do mainland species and ecosystems, and will examine the relationship between introduced species, decline/extinction of island species, and direct and indirect changes in ecosystem function as a result of invasion.

9.2 Are Islands Inherently More Subject to Invasion?

It is difficult to pinpoint the origin of the notion that islands are inherently more invasible than continental areas. However, this idea has been reinforced by the widely read publications of Carlquist (1965, 1974), who implied that island species were poor competitors and thus could not stand their ground against invaders. He stated: "continental species, steeled by competition, not only stand their ground, they are often preadapted to places far beyond the ranges they occupy.... Island creatures, reared in the hothouse like conditions of isolation, have no such reserves of rampant capabilities. Rather they often fall victim not only to man and the plants and animals he introduces, but... to the shortsighted qualities of their own evolution."

It has frequently been argued that islands are more subject to invasion because their native biota lack mechanisms to buffer them against change in the face of biological invasions (Carlquist 1965). The mechanisms that promote community "stability" in the face of biological invasions have been lumped under the general category of "biotic" or "ecological resistance" (e.g., Elton 1958; Simberloff 1986a; Brown 1989). The two major categories of interactions that are presumed to confer biotic resistance to mainland communities but not to islands are (1) well-established predator and/or pathogen populations, and (2) assemblages of competitors at or near some equilibrium density determined by resource availability. Island communities typically have low species richness and native island species show a high degree of specialization and endemism. This is suggested to result in abbreviated food web structure, undersaturated communities, lower competitive ability in island species compared to continental ones, and "vacant niche space".

The contention that islands inherently have low biotic resistance to invasion has not been well tested. There is limited evidence that species introductions have been hampered by generalist predators in both continental and island habitats. For example, Goeden and Louda (1976) examined cases of insect introductions for biological control purposes in both mainland and island habitats and found that in 15 out of 23 mainland cases, observational evidence existed that native predators or parasitoids were responsible for failed establishment of the introduced species. Predators also contributed to the failure of establishment on islands (five out of ten cases) but in three of these cases the predators were previously established introduced species. Few other studies are available from either mainland or island sites. Exceptions include D'Antonio (1993) and D'Antonio et al. (1993), who found that native generalist predators were important in suppressing invasion of California coastal plant communities by an aggressive non-native plant species, and Lake and O'Dowd (1991), who found that predation by native red land crabs limited invasion of pristine rainforest by the introduced African snail, *Achatina fulica*, on Christmas Island. This resistance to invasion in Christmas Island rainforests was not apparent in adjacent habitats, where the native vegetation had been disrupted by land clearing and was replaced by exotic plants. In those sites, native land crabs were rare and predation on marked *Achatina* was not detected. Kramer and D'Antonio (in prep.) found evidence that native land crabs on Moorea, Society Islands, limit invasion by the introduced mangrove, *Rhizophora stylosa*, into some intertidal habitats, although invasion is rapidly occurring into seemingly intact native habitats that naturally lack crabs.

The most common evidence used to support the contention that islands are more invasible than mainland sites are lists of the percent representation of alien species within particular taxonomic groups. Such evidence suffers from several shortcomings: (1) Lists rarely include data on the number of failed introductions. Hence, we do not know how many species fail to establish on islands because of native generalist predators or because of competition, so we cannot unambiguously compare the relative resistance of island and mainland sites to invasion. (2) Because islands generally have lower native species richness than comparable mainland areas, percentages may be higher on islands even when actual numbers of introduced species are lower. For example, continental islands off South Africa contain considerably fewer alien species than the adjacent mainland but a relatively higher percent of the total island flora is composed of aliens because the indigenous flora is very species poor (Cooper and Brooke 1986). Numbers of invading species rather than percentages might tell us more about "invasibility". Finally, (3) the data are not adjusted for area or habitat diversity which are both typically higher for continental sites compared to island sites. This is particularly problematic if mainland lists include areas away from a coastal influence. Interior habitats often have fewer invaders and have suffered less habitat disturbance than coastal areas (Table 1, Mooney et al. 1986). Coastal habitats, whether or not they are on islands, frequently harbor a large number and a high percentage of alien species. Likely factors promoting invasion in these areas include habitat disturbance due to urbanization and development, the benign environment of coastal sites, and high propagule influx due to proximity to ports and cities. Both numbers and percentages of alien species in the flora of the California Channel Islands are lower than in adjacent mainland regions but percentages are higher for the Channel Islands than for California as a whole (Table 9.1). Many coastal regions of South Africa have a higher number of introduced species than interior forested regions (MacDonald et al. 1986), suggesting that it is something about coastal environments, not necessarily islands, that make them likely to have a large number of introduced species.

Few studies exist on the role of competition in limiting invasions on islands. Moulton and Pimm (1986a, b), in their study of bird species introductions into Hawai'i, present data documenting both successes and failures of species introductions and the potential role of competition. However, their data suggest that it

Table 9.1. Alien plant representation in selected regions of California, USA. (M) indicates mainland sites, which precede the corresponding island sites. Data from Mooney et al. (1986) unless otherwise noted

Location	Total no. plant species	No. exotics	Exotics, %
Coastal California			
Santa Barbara vicinity (M)	1390	500	36.0
Santa Barbara Island	96	28	29.2
Northern Channel Islands[a]	334	79	23.6
Santa Monica Mts. (M)	805	206	25.6
Santa Catalina Island	559	106	19.0
San Francisco County (M)	1126	456	41.3
San Bruno Mountain (M)	401	117	29.2
Angel Island	406	124	30.5
Interior California			
Kern County	1713	257	14.6
Lassen National Park	715	37	5.2
Mt. Diablo	644	90	14.1
Mt. Hamilton Range	761	46	6.0
White Mountains	2280	137	5.9
California (entire state)[b]	5000+	650–750	13–15

[a] Halvorson (1992).
[b] Munz (1968).

is competition from previously established introduced birds that affects further invasion by introduced birds. Data on the interaction of native birds with introduced ones are lacking. Simberloff (1986a) explores the role of competition in limiting species establishment in island and mainland communities by examining cases of successful vs. failed insect introductions for biological control. He formulates two main arguments: (1) If mainland species are better competitors than island species or island habitats are more readily invaded, then mainland species introduced to island habitats should establish more frequently than island species introduced into mainland sites. The scant data do not support this contention. Secondly, if competition is important in limiting invasion onto islands, then insect families with higher native diversity on islands should be more resistant to invasion than families with few or no island representatives and vice versa. The data suggest that families with higher native diversity also tend to have higher numbers of introduced representatives. Roughgarden's experimental studies of lizards in the Bahamas suggest that intact lizard assemblages are resistant to further lizard species invasions and that competition can be important in limiting invasions onto islands (pers. comm.).

By comparing the flora of paired islands or island groups in which one member of the pair was disturbed by feral ungulates and the other was not, Merlin and Juvik (1992) concluded that alien plants establish more successfully on the grazed islands, and ungrazed island plant communities appear relatively resistant to invasion. Alien plants were present in ungrazed island settings examined; although propagules have indeed arrived and established there, they occupied a low percent cover or occurred only in disturbed microsites such as along trails.

A fundamental problem with studies of the causes of invasions on islands is that habitat alteration via direct human intervention or as a consequence of ungulate introductions, has been prevalent for so long on many islands that biotic resistance – if it existed – may have been eliminated by disturbance. This is suggested by the Lake and O'Dowd (1991) study of *Achatina* in undisturbed versus disturbed communities on Christmas Island. That habitat disturbance enhances the likelihood of invasion has been shown for continental habitats (Hobbs 1989) but it is often unclear what mechanisms promote invasion after disturbance. Disturbance may disrupt "biotic resistance" by eliminating predators and/or competitors or may cause a direct change in resource state (e.g., fire enhances N availability) and thereby favor exotic species.

There are purported examples of invasion by introduced plants into relatively undisturbed island habitats that support the idea that island plant communities are more easily invaded than mainland sites (Lorence and Sussman 1986; MacDonald et al. 1991; Stone et al. 1992). However, these studies concern recent invasions onto islands (Mauritius, Reunion, Hawaii) that have a long history of disturbance and have already lost substantial numbers of their native birds and plants and unknown numbers of invertebrates (e.g., Olson and James 1984; James, this Vol.). Thus, it may not be realistic to call these habitats undisturbed. Also, the "intact" native communities are often surrounded by severely degraded habitats that contribute alien propagules and increase the likelihood of alien establishment in the native habitat. Native species often recover after the elimination of disturbance on islands (e.g., Scowcroft and Conrad 1992), suggesting that they are competitive in the absence of introduced disturbance agents.

Oceanic islands typically have few native vertebrate species that might act as top predators or herbivores. This fact was noted by Darwin (1859), who stated: "I have not found a single instance...of a terrestrial mammal inhabiting an island more than 300 miles from a continent or great continental island...". This has clearly allowed successful invasion by vertebrates which face no predators or competitors and is consistent with the hypothesis that islands contain some "vacant niches" and are therefore more susceptible to invasion. Loope and Mueller-Dombois (1989) refer to this phenomenon as taxonomic disharmony, that is, the native biota of islands often lack representatives of some of the taxonomic groups common in continental regions. Many studies have documented the high degree of success of vertebrate invasions on islands (Moulton and Pimm 1986a,b; Brown 1989). For example, the island of Guam lacked a native snake fauna, and was readily invaded by the introduced brown tree snake, *Boiga irregularis* (Savidge 1987).

Ants are another ecologically important group of organisms that are often lacking in the native biota of isolated islands. The Hawaiian islands lack both native ants and mammals (with the exception of bats) and have been readily invaded by many species of each of these groups. Huddleston and Fluker (1968) report that over 40 species of ants have successfully established in Hawai'i.

With regard to "vacant niches", Loope and Mueller-Dombois (1989) make the observation that introduced species themselves frequently create new niches which are then exploited by other introduced species. For example, feral pigs severely disturb the understory of Hawaiian rainforests, creating a type of soil disturbance that did not previously occur. Disturbed patches of soil are then invaded by numerous introduced plants (Huenneke and Vitousek 1990; Aplet et al. 1991). Merlin and Juvik (1992) document the association of introduced plants with goat disturbances on several Pacific Islands. Likewise, introduced fruit-eating birds facilitate invasion of Hawaiian forests by the introduced fruit-bearing tree *Myrica faya* (Vitousek and Walker 1989).

A further line of reasoning that has been used as evidence that island species are poor competitors is that extinction among island species increases as a function of the percent of endemism (MacDonald et al. 1988; Loope and Mueller-Dombois 1989). This relationship has been interpreted to infer that endemic species are poorer competitors, and thus more prone to extinction in the face of habitat modification or disturbance and exotic encroachment. However, a similar pattern of high local endemism corresponding to high rates of extinction or population decline has been described for continental sites (Hall et al. 1984; Sukopp and Trepl 1987; Smith and Berg 1988). Such data appear more relevant to the question of why locally endemic species are more prone to extinction than to why island communities are more susceptible to invasion.

Overall it is difficult to find experimental support for the hypothesis that intact island communities inherently lack biotic resistance and are therefore relatively more susceptible to invasion. It is clear that oceanic islands often have underrepresented groups of species (disharmony), and it is likely that this has contributed to the success of many introduced species, particularly predators. However, many island communities were disturbed by indigenous peoples prior to the arrival of Europeans. Because islands have limited space, indigenous peoples on isolated islands could not readily move to new areas when site productivity began to deteriorate. Thus coastal lowland habitats on isolated islands were probably more severely impacted by pre-European peoples than were habitats on continents, and it is likely that many native plants and animals were driven to extinction before European arrival. This is well documented for bird species in Hawai'i and other Pacific Islands (Olson and James 1984; Steadman and Kirch 1990), including New Zealand (King 1984). Many such extinctions are thought to have resulted from human hunting pressure and habitat alteration rather than from the introduction of non-native species. A large number of extinctions probably also occurred after European contact and many potentially important species may have disappeared before scientific surveys were conducted. Thus, because of the historical and spatial extent of habitat disruption, the intrinsic ability of intact island communities to resist invaders cannot be rigorously tested.

9.3 Effects of Introduced Species

Introduced species can affect diversity of native species directly through species interactions such as predation, or indirectly by altering resource availability, turnover rates, or disturbance regime which in turn affect species composition.

9.3.1 Species-Level Effects

Although the precise causes of species declines and extinctions are often difficult to identify (Simberloff 1986b), it appears that island species are more subject to decline and extinction as a result of introduced species than are mainland species. For example, we surveyed the suspected causes of extinction or decline for federally listed bird species from the continental United States and Hawai'i. Alien species were listed as at least one of the causes of decline for 16 out of 19 Hawaiian bird species and only 2 out of 13 mainland species (Table 9.2). These data are consistent with, but even more extreme than those presented by MacDonald et al. (1989) for IUCN surveys of threatened vertebrate species from throughout the world. They found that 38%of endangered insular bird species are threatened by introduced species while only 5% of mainland species are so threatened. Of continental vertebrate species in the USA, we found that the only group overwhelmingly threatened by introduced species are freshwater fishes (Table 9.2): of 19 species listed as threatened or endangered in the original Endangered Species Act of 1967, 11 were threatened, at least in part, by introduced fishes. Recent records of amphibian declines suggest that exotic species such as trout and bullfrogs have caused local extinction of frogs throughout the western United States (Corn 1994). Corn (1994) lists six species directly threatened by these predators, but states that few data are available on the total number of declining anurans. Aquatic habitats such as streams and springs are more isolated from each other than are most terrestrial continental habitats and thus may represent ecological conditions that are more similar to geographically isolated islands. Brown (1989) came to a similar conclusion after surveying vertebrates in the southwestern United States.

While data on native plant assemblages are more difficult to obtain, information we could locate supports the idea that many endangered island species are threatened by exotic species. Of the 134 state-listed threatened or endangered plant species in continental California, exotic species (plant or animal) are listed

Table 9.2. Role of exotic species in the decline of US federally listed threatened and endangered species. (Wilcove et al. 1993)

	No. native species	No. threatened by exotics	Frequency threatened
Hawaii (birds)	19	16	84
US mainland (birds)	13	2	15
US mainland (fish)	19	11	58

as contributors to the decline of only 21 (16%), and in most cases exotics were not the sole cause of decline. By contrast, on the California Channel Islands, 17 of the 19 listed species have been reduced by feral animals, including rabbits, goats, pigs, and sheep (California Dept. Fish and Game 1990). These numbers are more extreme than those summarized by the Dept. of Fish and Game (1990) perhaps because the state did not consider sheep as feral animals. However, for threatened species on Santa Cruz Island, where sheep were abundant and feral for at least the last 60 years (van Vuren and Coblentz 1987), we considered "sheep" as feral animals rather than as "livestock", contributing to the extremely high percentage of island plant species threatened by feral animals. On the mainland, at least as many species are threatened by exotics as on islands, but other sources of threat such as urbanization and residential development exact a larger toll on native plants (Table 9.3).

It is clear that one class of organisms, vertebrates, is responsible for a large percent of the numerous examples of island species declines or extinctions. For threatened plant species, feral pigs, goats, and sheep trample and consume native species and reduce habitat suitability. Many birds, mammals, and lizards have been driven to extinction by predators such as cats, mongoose, snakes, and rats. Examples of extinctions or declines thought to be due to introduced species are listed in Table 9.4.

Recently, there has been a growing interest in extinctions of invertebrates, but little quantitative data exists on island invertebrate declines other than information on island snails (Hadfield 1986; Murray et al. 1988; Hadfield and Miller 1989). A major cause of invertebrate decline is introduction of invertebrate predators such as snails, wasps, and ants (Table 9.4). A recent study by Cole et al. (1992) documents the detrimental effect of the invading Argentine ant on invertebrates in Hawaiian shrublands. Data are not readily available to compare causes of invertebrate extinctions on mainland versus island habitats.

Table 9.3. Threats to California state listed threatened and endangered plant species, not including wetland habitats. Only the most common threats are listed, but a species can be threatened by several factors. (California Department of Fish and Game 1990)

	Channel Islands		California mainland	
No. of species	19		134	
Causes known	19		132	
Threats	No.	%	No.	%
Exotic species	17	89	21	16
Urbanization	0	–	69	51
Recreational development	0	–	48	36
Livestock grazing	4+	>21	38	28
Mining	0	–	20	15
Agriculture	0	–	23	17
Fire suppression	0	–	22	16

It is important to note that many of the invaders responsible for species declines and extinctions on islands are also present as aliens in mainland habitats, yet rarely are their effects as obvious in continental sites. For examples, black rats (*Rattus rattus*) and brown rats (*Rattus norwegicus*) are present in nature reserves in Africa, Australia, and North America (Loope et al. 1988; MacDonald and Frame 1988; MacDonald et al. 1988), yet have not become widespread predators or pests. Feral cats are also present in nature reserves throughout the world (e.g., MacDonald and Frame 1988; MacDonald et al. 1988) but have only been suspected of causing extinction in isolated reserves (MacDonald et al. 1988) or on islands. Feral pigs are a persistent understory disturbance agent in many continental habitats in North America, South Africa, and Australia (e.g., Wood and Barrett 1979; Oliver 1984) and while they are often considered a management concern, they are generally not associated with severe declines or extinctions of continental species. In the Smoky Mountains of eastern North America, feral pigs are known to be destructive to understory plant species but these species are not in danger of extinction (Bratton 1974, 1975), perhaps because these plant species are relatively widespread and not yet universally exposed to pig impacts.

Why do introduced and feral animals appear to cause more extinction or decline on islands compared to mainlands? It is possible that some biotic mechanism (i.e., competition or predation) is keeping predator populations in check on the mainland, although there are many places where such feral animals are reputed to be abundant (e.g., Oliver 1984). Thus, the relatively strong effect of introduced animals on native island species is probably due to demographic

Table 9.4. Examples of studies suggesting causative relationship between introduced predators and decline of native island species

Island(s)	Predator	Declining/extinct native species	Reference
Antilles	Rats	Rodents	Woods (1989)
Florida Keys	Rats	Silver rice rat	Goodyear (1987)
Guam	Brown tree snake	Forest birds	Savidge (1987) Engbring and Fritts (1988)
		Lizards	Rodda and Fritts (1992)
	Shrew	Gecko	Rodda (1992)
	Biocontrol insects	Butterflies	Nafus (1993)
Hawai'i	Rats	Forest birds	Atkinson (1977)
		Endemic rail	Greenway (1967)
	Cats, mongoose	Forest birds	Atkinson (1977)
	Ants	Arthropods	Howarth (1985)
Lord Howe Island	Rats	Birds	King (1984)
Newfoundland	Red squirrel	Red crossbill	Pimm (1990)
New Zealand	Rats, stoats, cats	Birds	King (1984)
Society Islands	*Euglandina*	*Partula* spp. (snails)	Murray et al. (1988)
West Indies	Cats, mongoose, dogs	Amphibians Reptiles	Henderson (1992)

characteristics of island species. Island species are by definition isolated. They often have limited powers of dispersal, less total suitable habitat, are comprised of fewer populations, and/or have a smaller total population size than mainland species. That small populations are more prone to extinction has been demonstrated by several investigations (Diamond 1984; Pimm 1989). Island species are less likely to function as metapopulations where immigration and gene flow among demes and recolonization can occur after local extinction. Also, if island species do have more limited dispersal than their mainland counterparts, effective population size of a series of demes will be lower and more rapid local and global extinction is likely. Wilson (1961) in his presentation of the taxon cycle concept, theorized that populations of island species become increasingly specialized and fragmented over time, and as this occurs their vulnerability to extinction increases.

9.3.2 Ecosystem-Level Effects

Invading species in any system can alter ecosystem properties if they (1) differ from native species in the way that they acquire and use resources, (2) alter trophic structure of an area, and/or (3) alter the disturbance regime (Vitousek 1990). Many invading species directly or indirectly modify systems by at least two of these mechanisms. For example, by utilizing salts in the soil differently from other species in the communities, the invading crystalline iceplant, *Mesembryanthemum crystallinum*, causes heavy deposition of salt on the soil surface (Vivrette and Muller 1977). This in turn alters soil nutrient processes. Salt deposited by *M. crystallinum* suppresses germination and growth of other plant species so that areas become monocultures of *Mesembryanthemum* (Vivrette and Muller 1977; Kloot 1983). This ultimately may alter trophic structure by eliminating animal species that cannot utilize iceplant for food or shelter.

It is clear that many species which have invaded islands can alter ecosystem-level processes by altering resource availability or rates of resource renewal. *Mesembryanthemum crystallinum* has heavily invaded large areas on the California Channel Islands. The nitrogen-fixing tree, *Myrica faya*, from the Canary islands is invading large portions of mesic and dry woodlands on the island of Hawai'i. This tree adds four times more nitrogen to the soil each year than can be fixed by native species (Vitousek et al. 1987). Such effects, however, are probably not unique to islands. The nitrogen-fixing shrub, *Cytisus scoparius* (scotchbroom) has invaded habitats throughout Washington, Oregon, and northern California, and the nitrogen-fixing tree *Casuarina* has invaded extensive areas of southern Florida (La Rosa et al. 1992). The ecosystem-level effects of these invasions have not been quantified, but could be large. Musil (1993) and Witkowski (1990) have shown increased soil inputs in continental South Africa as a consequence of invasion of these sites by the Australian N-fixing tree, *Acacia saligna*.

That introduced species can alter disturbance regimes has already been mentioned for the case of ungulates such as pigs. However, a now common alteration of disturbance regimes in tropical islands throughout the world is alteration of the fire cycle (D'Antonio and Vitousek 1992). For example, introduced grasses have promoted fire throughout the islands of Oceania by causing accumulation of fuel (D'Antonio and Vitousek 1992). Few native island species show adaptations to fire. As a result, the abundance and number of native species has decreased in communities experiencing fire. In seasonally dry submontane woodlands in Hawai'i, the alien grass *Schizachyrium condensatum* has promoted larger and more frequent fires (Smith and Tunison 1992). Richness and cover of native plant species has declined along transects in these sites (Table 9.5). In addition, fire has caused a loss in aboveground nitrogen and a change in the distribution of nitrogen pools within these ecosystems (D'Antonio and Vitousek, submitted). While this example clearly demonstrates the effect that invading species can have on communities and ecosystems, the feedback between changes in ecosystem processes and loss of native biological diversity is still unclear. Does the change in nitrogen pools that occurs in direct response to fire affect the ability of native species to recover? Does decreased biological diversity in direct response to fire result in changes in ecosystem processes in the recovering postfire community?

Changes in fire regimes in response to the presence of alien species are not unique to islands. Alien grass-fueled fires are now common in the Americas and Australia (D'Antonio and Vitousek 1992). *Bromus tectorum* (cheatgrass), a grass of European origin, now dominates thousands of hectares of land in western North America. Fire frequencies have greatly increased in these areas (Klemmendson and Smith 1964; Whisenant 1990) and native biological diversity has declined (Billings 1990). In Africa, fire promoted by nongrass exotic plants is beginning to penetrate riparian communities that were previously fire-resistant (MacDonald et al. 1988).

An ecosystem-level change resulting from exotics that does appear to be primarily associated with islands is alteration in trophic structure due to introduced predators such as cats, rats, or invertebrate predators. (Aquatic food webs on continents have also been dramatically altered by introduced predators.) Greenway (1967) documented avian extinctions in response to introduced predators and found that 90% of the examples of such extinctions come from islands. Still, the effects of these extinctions on energy flow, nutrient cycling, and other ecosystem processes have rarely been studied.

Table 9.5. Cover and richness of native species along transects in unburned woodland and woodland burned in alien-grass fueled fires in Hawaii Volcanoes National Park. Cover measured with point-intercept sampling and presented here as absolute cover; n = 10 transects per habitat

	Unburned	Once burned (20-year postfire)	Twice burned (4 and 20-year postfire)
Native species			
Mean No.	7.8	2.9	1.2
Range	7–10	1–3	1–2
Cover, %	155.3	51.2	6.2
(SE)	8.4	5.7	1.2

Few large-scale effects of introduced predators have been documented on continents, but alterations in dominance relationships due to introduced pathogens have occurred and these must be having or have had effects on trophic dynamics. Although introduced pathogens have been responsible for severe declines of island bird species in Hawai'i (Wilcove et al. 1993) and declines of ungulates in Africa (MacDonald and Frame 1988), their most spectacular effects have been on tree communities in temperate forests. The most dramatic examples include the loss of the once dominant American Chestnut tree, *Castanea dentata*, from the deciduous forests of the eastern United States in the early 1900s. It was replaced at least partially by oaks (*Quercus* spp.) after it was almost completely extirpated by the introduced fungus, *Endothia parasitica* (McCormick and Platt 1979). The functional equivalency of chestnut and the species which replaced it has not been evaluated. In the western United States and Canada, stands of western white pine (*Pinus monticola*) have been decimated by the Asian white pine blister rust, *Cronartium ribicola*. In parts of Idaho and Montana, white pine has declined from as much as 70% of the forest canopy to less than 6% (Hagle and Byler, unpubl. data), but again the ecosystem consequences of these changes have not been quantified. Pathogen outbreaks on forest trees have been reported for islands including Japan and Australia (von Broembsen 1989) but there is no obvious pattern of increased invasion or increased effects of plant pathogen invaders on islands.

9.4 Ecosystem Function and Species Loss As a Result of Invasions

While it is clear that species introductions both on islands and on continents can cause changes in ecosystem processes, it is less clear that species extinctions caused by invasion have changed ecosystem function. In most cases where dominant species have experienced reduced densities because of introduced predators or pathogens, there has been little or no follow-up assessment of the consequences of species loss to the community and to ecosystem function. Trophic structure may be dramatically altered by these changes but few studies actually document this. Even in the well-known cases of the decline of chestnut and elm (*Ulmus americana*) from American forests, although shifts in the dominance relationships of canopy trees have been documented (McCormick and Platt 1980; Huenneke 1983) and some ecosystem consequences have been speculated (Shugart and West 1977), there have been no published quantitative studies of the effects of species loss on trophic dynamics or the impacts on wildlife species that almost certainly utilized these trees.

Ecosystem impacts of species declines or extinctions are likewise little known in island systems. We would expect that species losses would alter ecosystem processes when major players are lost from a system. Most of the plant species in danger of extinction as a result of invasions on the California Channel Islands (Appendix 9.1), are small plants that are unlikely to have been community dominants. However, four taxa are nitrogen-fixing legumes and at least three of

Appendix 9.1. Threatened or endangered plant species on California Channel Island

Species	Life-form	Threats
Astragalus traskiae[a]	Perennial herb	Rabbits
Castilleja grisea	Perennial subshrub	Goats
Cercocarpus traskiae	Shrub	Pigs, deer
Delphinium variegatum ssp. *kinkiense*	Annual herb	Goats, pigs, exotic plants
Dudleya nesiotica	Perennial succulent	Sheep, pigs
Dudleya traskiae	Perennial succulent	Rabbits
Eriogonum grande ssp. *timorum*	Perennial mat	Goats, sheep
Eriogonum giganteum ssp. *compactum*	Shrub	Goats, rabbits
Galium buxifolium	Small shrub	Sheep
Galium catalinense ssp. *catalinense*	Shrub	Goats, pigs
Lithophragma maximum	Rhizomatous herb	Goats, pigs
Lotus argophyllus ssp. *adsurgens*[a]	Subshrub	Goats, sheep
Lotus argophyllus ssp. *niveus*[b]	Subshrub	Sheep, cattle, exotic plants
Lotus dendroides ssp. *traskiae*[a]	Subshrub	Pigs, goats
Berberis pinnata ssp. *insularis*	Shrub or vine	Sheep, cattle
Malacothrix clementina	Annual herb	Goats, pigs
Malacothamnus fasciculatus ssp. *nesioticus*	Shrub	Sheep, cattle

[a] Nitrogen-fixing taxa.

them have congeneric relatives that are quite common on the mainland. Decline of these taxa may already have had important consequences for invertebrate herbivores, soil microorganisms, and soil nutriets. They have been replaced by European grasses, annual herbs, or crystalline iceplant. None of the replacement species fixes nitrogen, and all have different phenology from the perennial legumes. Examination of the 111 federally listed endangered plants in the Hawaiian islands suggests that most of the listed species were unlikely to have been community dominants in recent history (Cuddihy, pers. comm.). Most of the species were probably more abundant in the past than today, so their potential contribution to ecosystem processes such as energy flow, trophic structure and biogeochemical cycling is difficult to discern. Several of these species are also nitrogen fixers, and may have played a more important role prior to European settlement.

It may be misleading to use endangered species as an assessment of the effects of invaders on anything more than biological diversity or species richness. Many endangered species probably were never major players at the ecosystem level, and evidence suggests that many are restricted endemics whether on islands or on continents. Although there are cases to the contrary, dominant species, even those greatly reduced in number over large areas, do not often end up on endangered species lists. Their distributions are likely to be primarily altered by land-use change in conjunction with alien species invasions. In Hawai'i, *Erythrina sandwicense* is thought to have been one of the dominant native trees in some arid lowland environments before the arrival of humans (Cuddihy and Stone 1990). As a result of land-use change and exotic invasions, it and other possible codominants, now uncommon but still in lowland habitats, have been replaced by the introduced nitrogen-fixing tree, *Prosopis pallida*. *Erythrina* is not on any "endangered species list", and no information is available on what this total change in dominance has meant to ecosystem processes in these habitats. Alien grass-fueled fire in Hawai'i has caused the elimination of *Metrosideros polymorpha*, the dominant canopy tree, and *Styphelia tamiameae*, the dominant understory shrub from small patches of dry forest. Clearly these local-scale species losses cause a change in trophic structure – but the effect of their loss on rates of biogeochemical cycling or other ecosystem processes has yet to be determined.

9.5 Conclusions

Regardless of the causes of invasion onto islands, it is clear that many islands harbor a large number of introduced species. These introduced species can directly impact biological diversity in two ways: (1) if no species disappear in response to the invader, exotics will increase the number of species in the system. This appears to be the most common result of successful invasion (Simberloff 1981), although the impacts of an invader may be initially subtle and not fully manifested for a long period; (2) conversely, exotic species may decrease biological diversity by eliminating or reducing native species through competition, predation or disease.

It is clear that species additions to communities can have tremendous effects at all levels of biological organization. Changes in ecosystem processes and hence ecosystem function should be expected when introduced species utilize resources differently from the existing species and thus alter rates of resource supply, or when they alter the disturbance regime and/or the trophic structure. Examples such as *Myrica faya* invasion into Hawaiian woodlands or alien grass invasions into arid or semi-arid habitats clearly demonstrate large-scale single species effects, and case studies from both islands and continental habitats are becoming increasingly common. Only direct changes in trophic structure clearly occur more often or with greater severity of effects on islands than on continents because islands typically have poorly represented groups of organisms, particularly predators and large herbivores, and native island species appear to be more prone to decimation than most mainland species.

The higher level effects of species losses, particularly losses that result from invasion, are less well documented than species additions. Losses would be expected to affect trophic dynamics if the species that replace those lost do not provide the same sort of food or shelter for other organisms in the communities or do not influence rates of nutrient and energy flow in the same way. It is only recently that we have begun to identify the precise causes of species extinctions (Simberloff 1986b) and we do know that species introductions have caused extinction of island species. However, there are almost no studies of the effect of these extinctions on ecosystem processes. Thus, we cannot yet link introduced species to extinctions and subsequent changes in ecosystem function. However, because of the generally low species richness on islands and the discrete nature of island habitats, it may be on islands rather than on continents that we will unequivocally be able to make these linkages.

Acknowledgments. We would like to thank the participants in the SCOPE Conference on Islands and Ecosystem Function for their productive feedback, and the Bahamian National Trust for facilitating this conference. We also thank P. Vitousek and L. Loope for their useful comments on drafts of this manuscript. Flint Hughes assisted in collection of data presented in Table 9.5. For financial support, we acknowledge the W. Alton Jones Foundation for its grant to the Pacific Institute, and National Science Foundation Grant No. BSR-91-19618 to C. D'Antonio.

References

Aplet G, Anderson S, Stone CP (1991) Association between feral pig disturbance and the composition of some alien plant assemblages in Hawaii Volcanoes National Park. Vegetatio 95: 55–62

Atkinson IAE (1977) A reassessment of factors, particularly *Rattus rattus* L., that influenced the decline of endemic forest birds in the Hawaiian Islands. Pac Sci 31: 109–133

Billings W (1990) *Bromus tectorum*, a biotic cause of ecosystem impoverishment in the Great Basin. In: Woodell M (ed) The Earth in transition: patterns and processes of biotic impoverishments. Cambridge University Press, Cambridge, pp 301–322

Bratton SP (1974) The effect of the European wild boar (*Sus scrofa*) on high-elevation vernal flora in the Great Smoky Mountains National Park. Bull Torrey Bot Club 101: 198–206

Bratton SP (1975) The effect of the European wild boar (*Sus scrofa*) on grey beech forest in the Great Smoky Mountains. Ecology 56: 1356–1366

Brockie RE, Loope LL, Usher MB, Hamann O (1988) Biological invasions of island nature reserves. Biol Conserv 44: 9–36

Brown JH (1989) Patterns, modes and extents of invasions by vertebrates. In: Drake JA, Mooney H, di Castri F, Groves R, Kruger F, Rejmanek M, Williamson M (eds) Biological invasions: a global perspective. Wiley, New York, pp 85–109

California Department of Fish and Game (1990) Report on the status of state-listed threatened and endangered animals and plants. California Resources Agency, Sacromento

Carlquist S (1965) Island biology: a natural history of the islands of the world. Natural History Press, Garden City, NJ

Carlquist S (1974) Island biology. Columbia University Press, New York, fauna of Hawaiian high-elevation shrubland. Ecology 73: 1313–1322

Cole FR, Madeiros AC, Loope LL, Zuehlke WW (1992) Effects of the Argentine ant on arthropod fauna of Hawaiian high-elevation shrubland. Ecology 73: 1313–1322

Cooper J, Brooke RK (1986) Alien plants and animals on South African continental and oceanic islands: species richness, ecological impacts and management. In: MacDonald IA, Kruger FJ, Ferrar AA (eds) The ecology and management of biological invasions in southern Africa. Oxford University Press, Oxford

Corn PS (1994) What we know and don't know about amphibian declines in the West. In: Covington WW, DeBano LF (tech. coord.) Sustainable ecological systems: implementing an ecological approach to land management. USDA–Forest Service GTR RM-247, Fort Collins, Colorado, pp 59–67

Cuddihy L, Stone CP (1990) Alteration of native Hawaiian vegetation: effects of humans, their activities and introductions. University of Hawaii Press, Honolulu

D'Antonio CM (1993) Mechanisms limiting invasion of coastal plant communities by the alien succulent *Carpobrotus edulis*. Ecology 74: 83–95

D'Antonio CM, Vitousek PM (1992) Biological invasions by exotic grasses, the grass/fire cycle and global change. Annu Rev Ecol Syst 23: 63–88
D'Antonio CM, Odion D, Tyler C (1993) Invasion of maritime chaparral by the introduced succulent *Carpobrotus edulis*: the roles of fire and herbivory. Oecologia 95: 14–21
Darwin C (1859) On the origin of species. Penguin Books, Harmondsworth
Diamond JM (1984) Historic extinctions: a Rosetta Stone for understanding prehistoric extinctions. In: Martin PS, Klein RG (eds) Quaternary extinctions: a prehistoric revolution. University of Arizona Press, Tucson, pp 824–862
Diamond J, Case T (1986) Overview: introductions, extinctions, exterminations and invasions. In: Diamond J, Case T (eds) Community ecology. Harper and Row, New York, pp 65–79
Drake JA, Mooney H, di Castri F, Groves R, Kruger F, Rejmanek M, Williamson M (eds) (1989) Biological invasions: a global perspective. Wiley, New York
Elton C (1958) The Ecology of invasions by animals and plants. Chapman and Hall, New York
Engbring J, Fritts TH (1988) Demise of an insular avifauna: the brown tree snake on Guam. Trans West Sect Wildl Soc 24: 31–37
Goeden R, Louda S (1976) Biotic interference with insects imported for weed control. Annu Rev Entomol 21: 325–342
Goodyear NC (1987) Distribution and habitat of the silver rice rat, *Oryzomys argentatus*. J Mamm 68: 692–695
Goodyear N (1992) Spatial overlap and dietary selection of native rice rats and exotic black rats. J Mammal 73: 186–200
Greenway C Jr (1967) Extinct and vanishing birds of the world. Dover, New York
Hadfield M (1986) Extinction in Hawaiian achatinelline snails. Malacologia 27: 67–81
Hadfield M, Miller S (1989) Demographic studies on Hawaii's endangered tree snails: *Partulina proxima*. Pac Sci 43: 1–16
Hall AV, de Winter B, Fourie S, Arnold TH (1984) Threatened plants in Southern Africa. Biol Conserv 28: 5–20
Halvorson W (1992) Alien plants at Channel Island National Park. In: Stone CP, Smith CW, Tunison JT (eds) Alien plant invasions in native ecosystems of Hawai'i: management and research. University of Hawaii Press, Honolulu, pp 64–96
Henderson RW (1992) Consequences of predator introductions and habitat destruction on amphibians and reptiles in the post-Columbus West Indies. Carribean J Sci 28: 1–10
Hobbs R (1989) The nature and effects of disturbance relative to invasions. In: Drake JA, Mooney H, di Castri F, Groves R, Kruger F, Rejmanek M, Williamson M (eds) Biological invasions: a global perspective. Wiley, New York, pp 389–405
Howarth FG (1985) Impacts of alien land arthropods and mollusks on native plants and animals in Hawaii. In: Stone CP, Scott JM (eds) Hawaii's terrestrial ecosystems: preservation and management. University of Hawaii Press, Honolulu, pp 149–179
Huddleston EW, Fluker SS (1968) Distribution of ant species in Hawaii. Proc Hawaii Entomol Soc 20: 45–69
Huenneke L (1983) Understory response of gaps caused by the death of *Ulmus americana* in central New York. Bull Torrey Bot Club 110: 170–175
Huenneke L, Vitousek P (1990) Seedling and clonal recruitment of the invasive tree *Psidium cattleianum*: Implications for management of native Hawaiian forests. Biol Conserv 53: 199–212
King C (1984) Immigrant killers – introduced predators and the conservation of birds in New Zealand. Oxford University Press, Oxford
Klemmendson JO, Smith JG (1964) Cheatgrass (*Bromus tectorum* L.). Bot Rev 30: 226–261
Kloot PM (1983) The role of common iceplant (*Mesembryanthemum crystallinum*) in the deterioration of medic pastures. Aust J Ecol 8: 301–306
Lake S, O'Dowd DJ (1991) Red crabs in rain forest, Christmas Island: biotic resistance to invasion by an exotic snail. Oikos 62: 25–29
La Rosa AM, Doren RF, Gunderson L (1992) Alien plant management in Everglades National Park: an historical perspective. In: Stone CP, Smith CW, Tunison, JT (eds) Alien plant invasions in native ecosystems of Hawai'i: management and research. University of Hawaii Press, Honolulu, pp 47–63

Loope L, Mueller-Dombois D (1989) Characteristics of invaded islands, with special reference to Hawai'i. In: drake JA, Mooney H, di Castri F, Groves R, Kruger F, Rejmanek M, Williamson M (eds) Biological invasions: a global perspective. Wiley, New York, pp 257–281
Loope L, Sanchez P, Tarr P, Loope W, Anderson R (1988) Biological invasion of arid land nature reserves. Biol Conserv 44: 95–118
Lorence D, Sussman R (1986) Exotic species invasion into Mauritius wet forest remnants. J Trop Ecol 2: 147–162
MacDonald I, Frame GW (1988) The invasion of introduced species into nature reserves in tropical savannas and dry woodlands. Biol Conserv 44: 67–93
MacDonald I, Graber D, DeBenedetti S, Groves R, Fuentes E (1988) Introduced species in nature reserves in Mediterranean-type climatic regions of the world. Biol Conserv 44: 37–66
MacDonald IA, Loope LL, Usher M, Hamann O (1989) Wildlife conservation and the invasion of nature reserves by introduced species: a global perspective. In: Drake JA, Mooney H, di Castri F, Groves R, Kruger F, Rejmanek M, Williamson M (eds) Biological invasions: a global perspective. Wiley, New York
MacDonald IA, Thebaud C, Strahm W, Strasberg D (1991) Effects of alien plant invasions on native vegetation remnants on La Reunion (Mascarene Islands, Indian Ocean). Environ Conserv 18: 51–61
MacDonald IAW, Kruger FJ, Ferrar AA (1986) The ecology and management of biological invasions in southern Africa. Oxford University Press, Oxford
McCormick JF, Platt R (1980) Recovery of an Appalachian forest following the chestnut blight or Catherine Keever – you were right. Am Midl Nat 104: 264–273
Merlin M, Juvik J (1992) Relationships among native and alien plants on Pacific Islands with and without significant human disturbance and feral ungulates. In: Stone CP, Smith CW, Tunison JT (eds) Alien plant invasions in native ecosystems of Hawai'i: management and research. University of Hawaii Press, Honolulu, pp 597–624
Monnig E, Byler J (1992) Forest health and ecological integrity in the northern Rockies. USDA Forest Serv, FPM Report 92-7, Ft Collins, Colorado
Mooney H, Hamburg SP, Drake JA (1986) The invasions of plants and animals into California. In: Mooney HA, Drake J (eds) Ecology of biological invaions of North America and Hawaii. Springer, Berlin Heidelberg New York, pp 250–269
Moulton M, Pimm S (1986a) The extent of competition in shaping an introduced avifauna. In: Diamond J, Case T (eds) Community ecology. Harper and Row, New York, pp 80–97
Moulton M, Pimm S (1986b) Species introductions to Hawai'i. In: Mooney HA, Drake J (eds) Ecology of biological invasions of North America and Hawaii. Springer, Berlin Heidelberg New York, pp 231–249
Munz PA (1968) Supplement to *A California Flora.* University of California Press, Berkeley
Musil CF (1993) Effect of invasive Australian acacias on the regeneration, growth and nutrient chemistry of South African lowland fynbos. J Appl Ecol 30: 361–372
Murray J, Murray J, Johnson M, Clarke B (1988) The extinction of *Partula* on Moorea. Pac Sci 42: 150–153
Nafus DM (1993) Movement of introduced biological control agents onto nontarget butterflies, *Hypolimnas* spp. (Lepidoptera: Nymphalidae). Environ Entomol 22: 265–272
Oliver WLR (1984) Introduced feral pigs. In: Munton PN, Clutton-Brock J, Rudge MR (eds) Feral mammals – problems and potential. IUCNNR, 3rd Int Theriological Conf
Olson S, James H (1984) The role of Polynesians in the extinction of the avifauna of the Hawaiian Islands. In: Martin PS, Klein RG (eds) Quaternary extinctions: a prehistoric revolution. University of Arizone Press, Tucson, pp 768–780
Pimm S (1990) The decline of the Newfoundland crossbill. TREE 5: 350–351
Pimm SL (1989) Theories of predicting success and impact of introduced species. In: Drake JA, Mooney H, di Castri F, Groves R, Kruger F, Rejmanek M, Williamson M (eds) Biological invasions: a global perspective. Wiley, New York, pp 351–367
Richardson AM (1992) Altitudinal distribution of native and alien landhoppers (Amphipoda: Talitridae) in the Ko'olau Range, O'ahu, Hawaiian Islands. J Nat Hist 26: 339–352
Rodda G (1992) Loss of native reptiles associated with introductions of exotics in the Marianna Islands. Pac Sci 46: 399–400

Rodda G, Fritts T (1992) The impact of the introduction of the colubrid snake *Boiga irregularis* on Guam's Lizards. J Herpetoe 26: 166–174
Savidge J (1987) Extinction of an island forest avifauna by an introduced snake. Ecology 68: 660–668
Scowcroft PG, Conrad CE (1992) Alien and native plant response to release from feral sheep browsing on Mauna Kea. In: Stone CP, Smith CW, Tunison JT (eds) Alien plant invasions in native ecosystems of Hawai'i: management and research. University of Hawaii Press, Honolulu, pp 625–665
Shugart HH Jr, West DC (1977) Development of an Appalachian deciduous forest succession model and its application to assessment of the impact of the chestnut blight. J Environ Manage 5: 161–179
Simberloff D (1981) Community effects of introduced species. In: Nitecki MMH (ed) Biotic crises in ecological and evolutionary time. Academic Press, New York, pp 53–83
Simberloff D (1986a) Introduced insects: a biogeographic and systematic perspective. In: Mooney HA, Drake J (eds) Ecology of biological invasions of North America and Hawaii. Springer, Berlin Heidelberg New York, pp 3–26
Simberloff D (1986b) The proximate causes of extinction. In: Raup D, Jablonski D (eds) Patterns and processes in the history of life. Springer, Berlin Heidelberg, New York, pp 259–276
Smith CW, Tunison JT (1992) Fire and alien plants in Hawai'i: Research and management implications for native ecosystems. In: Stone CP, Smith CW, Tunison JT (eds) Alien plant invasions in native ecosystems of Hawai'i: management and research. University of Hawaii Press, Honolulu, pp 394–408
Smith JP, Berg K (eds) (1988) Inventory of rare and endangered vascular plants of California. Califorina Native Plant Society, Sacramento
Steadman D, Kirch P (1990) Prehistoric extinction of birds on Mangaia, Cook Islands, Polynesia Proc Natl Acad Sci 87: 9605–9609
Stone C, Cuddihy L, Tunison T (1992) Responses of Hawaiian ecosystems to removal of feral pigs and goats. In: Stone CP, Smith CW, Tunison JT (eds) Alien plant invasions in native ecosystems of Hawai'i: management and research. University of Hawaii Press, Honolulu, pp 666–704
Sukopp H, Trepl L (1987) Extinction and naturalization of plant species as related to ecosystem structure and function. In: Schultze ED, Zwolfer H (eds) Potentials and limitations of ecosystem analysis, Springer, Berlin Heidelberg New York, pp 245–276
Usher M (1988) Biological invasions of nature reserves: a search for generalisations. Biol Conserv 44: 119–135
Van Vuren D, Coblentz B (1987) Some ecological effects of feral sheep on Santa Cruz Island, California, USA. Biol Conserv 41: 253–268
Vitousek P (1990) Biological invasions and ecosystem processes: towards an integration of population biology and ecosystem studies. Oikos 57: 7–13
Vitousek P, Walker L (1989) Biological invasion by *Myrica faya* in Hawai'i: plant demography, nitrogen fixation, ecosystem effects. Ecol Monogr 59: 247–265
Vitousek P, Walker L, Whiteaker L, Mueller-Dombois D, Matson P (1987) Biological invasion by *Myrica faya* alters ecosystem development in Hawaii. Science 238: 802–804
Vivrette N, Muller CH (1977) Mechanism of invasion and dominance of coastal grassland by *Mesembryanthemum crystallinum*. Ecol Monogr 47: 301–318
Von Broembsen SL (1989) Invasions of natural ecosystems by plant pathogens. In: Drake JA, Mooney H, di Castri F, Groves R, Kruger R, Rejmanek M, Williamson M (eds) Biological invasions: a global perspective. Wiley, New York, pp 77–83
Whisenant S (1990) Changing fire frequencies on Idaho's Snake River plains: ecological and management implications. In: Proc symp on cheatgrass invasion, shrub dieoff and other aspects of shrub biology and management. USFS GTR-INT 276, Intermountain Forest and Range Experiment Station, Ogden, Utah, pp 4–10
Wilcove D, Willcox S, Phillips A, Bean M (1993) Whatever happened to the class of' 67?: a report commemorating the 26th anniversary of the first endangered species list. Environmental Defense Fund Press, Washington, DC

Wilson EO (1961) The nature of the taxon cycle in the Melanesian ant fauna. Am Nat 95: 169–193
Witkowski ETF (1990) Effects of invasive alien acacia on nutrient cycling in the coastal low- lands of the Cape fynbos. J Appl Ecol 28: 1–15
Wood G, Barrett R (1979) Status of wild pigs in the United States. Wild Soc Bull 7: 237–246
Woods CA (1989) Endemic rodents of the West Indies: the end of a splendid isolation. In: Lidicker WZ (ed) Rodents: a world survey of species of conservation concern. Occas Pap IUCN Species Survival Comm 4: 1–60

10 Climate Change and Island Biological Diversity

L. L. LOOPE

10.1 Introduction

Peters and Darling (1985), in a pioneering paper on *The Greenhouse Effect and Nature Reserves*, singled out island species as a special case of geographically restricted species which are particularly vulnerable to climate change – if the latitudinal or altitudinal migration required by the climate change exceeds the limits of the island, extinction necessarily follows. They recognized a mitigating circumstance, however, in that climatic changes on oceanic islands might be relatively mild because the sea would moderate the air temperature change.

Additional problems for modern island species in the face of climate change are that (1) humans and their invasive introductions have eroded the already small populations of island species and have eliminated important components of biodiversity (James, Chap. 8, this Vol.; D'Antonio and Dudley, Chap. 9, this Vol.), and (2) when island species are forced by changing climates to migrate, it is likely that invasive introduced species will be able to migrate more rapidly and displace natives from suitable habitat. Without climate change, the ultimate fate of island species is problematic but perhaps hopeful, given a certain degree of resilience of island biota (Loope and Mueller-Dombois 1989) and increasing efforts of island nature reserve managers to intervene in the struggle between native organisms and invaders (Brockie et al. 1988). With substantial climate change, however, the likelihood of retaining some semblance of pristine island biological diversity appears dim.

This Chapter attempts to review what evidence there is of past climates of islands, with the aim of shedding some light on the magnitude of climate change to be expected from anthropogenically induced global warming and therefore the prospects for retaining meaningful remnants of island biological diversity over the next centuries. I base my approach on the generalization of Webb (1992) that changes in the globally averaged climate since the climax of the most recent full-glacial period have been similar in magnitude (ca. 4 ± 1 °C) to changes predicted for global warming induced by greenhouse gases (e.g., Schneider et al. 1992). I thus make the bold (and risky) assumption that island climates are likely to change in the near future no more than they have changed since the most recent full-glacial.

National Biological Survey, Haleakala National Park Field Station, P.O. Box 369, Makawao, Maui, Hawaii 96768, USA

Ecological Studies, Vol. 115
Vitousek et al. (eds) Islands

I focus on middle- and low-latitude oceanic islands, eliminating from consideration islands of high latitudes and continental islands. Biotas of most high-latitude islands were heavily affected by glaciation and land-bridge connections and usually bear a close resemblance to adjacent continental biotas. Biotas of middle- and low-latitude continental islands are likewise tied to the fate of continents; their environments are generally less buffered from climatic changes than are oceanic islands. Likewise, atolls and other low islands are eliminated from consideration here, since the potential effects of global warming are obvious. This type of island was much larger than now, when sea level was about 60 m lower during the last full glacial period (18 000 years ago). If all glaciers and ice sheets on earth melted, sea level would be raised about 30 m, completely inundating these islands and obliterating their biological diversity (unless coral growth rates can keep up, a possibility discussed by Smith and Buddemeir 1992). Their biological diversity would be significantly decreased by the 10–100-cm rise in sea level predicted by 2050 A.D. (Schneider et al. 1992).

10.2 Quaternary Climates of Islands: What Do We Know?

Relatively little is known of island paleoenvironments, although it is possible to piece together a tentative picture from work in a few islands and island groups. This situation is likely to improve as a result of ongoing and incipient research. Worldwide, research on climate change has been tremendously stimulated by concerns over greenhouse-induced global warming, and some findings will be applicable to islands.

Some of the factors that have influenced the long-term periodic changes in global mean temperature and regional climate fluctuations during the Late Quaternary are beginning to become understood (Webb 1992). Long-term rhythmic variations in the orientation of the earth's axis and its orbit about the sun which appear to be important include: (1) tilt of the earth's axis (which varies from 21.5° to 24.5° with a periodicity of 41 000 years); (2) precession of the equinoxes, which involves variation in how close the earth is to the sun at a given time, with a peridocity of 21 000 years; and (3) change in the shape (eccentricity) of the earth's orbit, with a periodicity of 100 000 years. These orbital changes alter the amount of radiation received seasonally, thus altering seasonality (the relative severity of summers and winters), and influencing circulation of the atmosphere and oceans. How orbitally caused changes in seasonality affect climates of oceanic islands is not yet clear, but effects undoubtedly occur.

10.2.1 Quaternary Environments of Oceans

Air temperatures of islands are strongly influenced by ocean temperatures. In the Hawaiian Islands, for example, the seasonal extremes in sea surface temperatures range from 22 °C in March to 26 °C in October; the extremes in average monthly mean air temperatures at sea level coincide with these values (Gavenda 1992), and in fact extremes in monthly mean air temperatures at high elevations vary by only about 4–6 °C. Paleoclimatologists have estimated past sea surface temperatures based on the species composition of marine plankton, primarily foraminifera, in ocean sediment cores (CLIMAP 1976). Ocean temperatures underwent large changes at high latitudes during the Late Quaternary; changes may have been much less pronounced in tropical oceans. For example, during the last full-glacial period, the generally accepted view is that sea surface temperatures varied geographically from being more than 10 °C lower than those today in the North Atlantic Ocean to being 2 °C higher than those today in parts of the tropical Pacific Ocean (Webb 1992). Low and middle latitude sea surface temperatures apparently underwent no more than about 2 °C of temperature depression during glacial periods, according to CLIMAP (1976).

The CLIMAP conclusions have been widely accepted, but they have recently come under strong challenges (Emiliani 1992; Colinavaux 1993). Two lines of evidence seem to contradict the CLIMAP findings and support the occurence of relatively great tropical sea-surface temperature cooling during glacial periods: (1) pollen analyses which seem to show large shifts in vegetation zones for tropical areas (e.g., Rind and Peteet 1985); and (2) paleotemperature records in coral reefs (^{18}O) (Aharon 1983) and in groundwater (noble gases) (Stute et al. 1992) which suggest substantial cooling. Emiliani's (1992) analysis suggests an area weighted average glacial-interglacial temperature of 5.0 °C for the tropical-subtropical belt, with 5.6–7.8 °C for the Caribbean and 2.6–3.6 °C in the tropical pacific.

10.2.2 Pertinent Information on Quaternary Environments of the African and South American Tropics

Research in recent decades has dislodged a traditional view that tropical forests are extremely rich in biological diversity because they were little affected by Quaternary climatic fluctuations. Pollen studies in lowland tropical forests in Africa and South America indicate an incidence of montane elements in lowland forests during glacial periods and enhanced seasonality during the early to middle Holocene (Servant et al.1993). This evidence of cooling could be reconciled with only a 1–2 °C lower sea surface temperature in tropical oceans during full glacial by the assumption that polar air masses reached tropical regions more frequently, bringing periodically low temperatures. The increased seasonality during the early to middle Holocene is attributed to a latitudinal shift in the Intertropical Convergence Zone (ITCZ) and by shortterm (orbitally induced?) changes in the zonal atmospheric circulation (Servant et al. 1993).

In contrast, Colinvaux (1993) interprets the pollen record as requiring cooling over the Amazon of 6–9 °C , which he believes is consistent with a lowering of altitudinal vegetation zones in Columbia by 1500 m and a 1500 m descent of mountain glaciers during the ice ages.

10.2.3 Hawaiian Islands

Relatively great detail is devoted here to the Hawaiian Islands because (1) more pertinent information is available for them than for most other island groups, and (2) they possess substantial complexity which illustrates the difficulty of making simple predictions of the consequences of global warming.

10.2.3.1 Present-Day Conditions

The present-day Hawaiian Islands have a remarkable diversity of climates. Over the open sea near Hawaii, rainfall averages 63–75 cm/year. Yet certain localities on the islands receive up to 15 times this amount; others receive less than one-third of it. The cause of this variability and of annual totals which can be among the greatest on earth, is principally the orographic rains which form within the northeasterly moist trade wind air as it moves in from the sea and overrides the steep and high terrain of the islands (Blumenstock and Price 1967). Hawaii's location in relation to the relatively stable Pacific High or anticyclone is a crucial determinant of the precipitation regime. When the Pacific High moves, as often happens during El Niño/Southern Oscillation (ENSO) events (Rasmussen and Wallace 1983), Hawaii's precipitation is sharply reduced (Lyons 1982; Chu 1989).

Another important climatic phenomenon associated with the Pacific High, as well as with other subtropical high-pressure cells elsewhere, is the trade wind inversion (TWI). This temperature inversion, present 70% of the time at 1200–2400 m altitude in Hawaii (Blumenstock and Price 1967) but most frequently at 1800–2400 (Giambelluca and Nullet 1991), comprises a major climatic discontinuity; above the TWI in Hawaii, incident solar radiation is very high, relative humidity is generally low, and rainfall is moderate (2500 mm/year or less); below the TWI, incident solar radiation is moderate because of normally high cloud cover, relative humidity is generally high, and rainfall is heavy (with amounts of up to 5000–10 000 mm/year) on windward slopes (Giambelluca and Nullet 1991; Giambelluca et al. 1986).

In present-day Hawaii, temperature decreases by an average of 0.5 °C per 100 m increase in elevation; whereas mean annual temperature at sea level is about 25 °C, mean annual temperature at the 4200 m summit of Mauna Kea is about 4 °C (Price 1983). In the vicinity of the TWI (1300–2400 m altitude), the average temperature gradient is less steep than the Hawaiian average.

10.2.3.2 What Do We know About Quaternary Paleoenvironments in Hawaii?

Temperature. Based on species composition of marine plankton in ocean sediment cores (CLIMAP 1976), August sea surface temperatures near Hawaii during the last glacial maximum have been estimated to be about 1–2 °C cooler than now. It is likely that polar air masses reached Hawaii more often during full glacial conditions than now, much as hypothesized for the tropics of Africa and South America by Servant et al. (1993). Currently, as many as 20 cold air mases reach Kauai (22 °N) each winter, but only half reach the island of Hawaii (20 °N) (Price 1993).

Wind Direction and Velocity. Evidence in the form of lithified sand dunes and asymmetrical morphology of cinder cones suggests that the prevailing wind direction (northeasterly trade winds) during glacial periods in the Hawaiian Islands was the same as today; there appears to be no evidence suggesting that prevailing wind direction has ever differed from now. Grain size of pelagic eolian sediments suggests a general correlation between glacial maxima and increased wind velocities (Chuey et al. 1987).

Precipitation. A computer simulation model of tropical climate during a glacial period (Manabe and Hahn 1977) predicted that mean rainfall over oceans at 5–30. N latitude would be greater during glacial conditions than now, and sparse pedological evidence suggests that rainfall may indeed have been greater at low elevations in the Hawaiian Islands during glacial periods than during interglacials (Gavenda 1992).

Trade Wind Inversion Level. According to the interpretation of Selling (1948), evidence of elevational shifts in the trade wind inversion (TWI) in the recent past are contained in the pollen record of montane mires on the island of Maui in the Hawaiian Islands. Although correlations with worldwide climatic events were hampered by lack of opportunity for radiocarbon dating at that time, Selling hypothesized that since the end of the last glacial period the TWI (now averaging about 2000 m altitude) was depressed to an average of about 1500 m during cool periods, and rose above the current levels during warm periods. During the periods when the TWI was depressed, the summit of West Maui's Puu Kukui (elevation 1760 m), which currently receives mean annual precipitation of nearly 10 000 mm, supported vegetation that is now typical of dry sites.

Glaciation of Mauna Kea. The glacial moraines on the volcanic peak of Mauna Kea, elevation 4206 m, at 20 °N on the island of Hawaii, provide what is probably the most dramatic evidence of Quaternary climate change for remote oceanic islands within the tropics or subtropics. Four glacial stages, spanning about 160 000 years, have been delineated and rather precisely dated (Dorn et al. 1991; Porter 1979). The most recent glacier, the Makanaka ice cap, was about 70 km^2 in area and 100 m thick and lasted from 40 000 years B.P. to 13 000 B.P.

The precise combination of factors that led to formation and growth of icecaps on Mauna Kea is far from clear. If snowline depression (about 935 m for the most recent glaciation) was only due to reduction of air temperature, and lapse rates were similar to those today, then July temperatures near Mauna Kea's summit would have had to be about 5 °C lower than today for glaciers to form (Porter 1979). (Gavenda (1992) raises the possibility that the adiabatic lapse rate may have been steeper than today, which would allow the summit to be 5 °C cooler whereas the ocean was only 1–2 °C cooler, based on a model by Potter et al. 1975). If mean air temperature on the upper slopes of the mountain were reduced only 1–2 °C, then a large increase in precipitation would be required. (Current mean annual precipitation near the Mauna Kea summit is only 40–50 cm). The Makanaka ice cap was remarkably symmetrical around the summit, with a bulge in the direction of the trade winds, suggesting that accumulation was the dominant control on snowline altitude rather than ablation. (If ablation were dominant, one would expect the snowline to be much higher on south-facing slopes than on north-facing slopes). Porter (1979) speculated that if precipitation did increase substantially during glaciation, it might have resulted from a weakening of the Hadley circulation and a corresponding weakening of the trade wind inversion over Hawaii, thereby permitting a greater percentage of the moisture carried by trade winds to reach the upper slopes of the mountain. Snowfall could also have been enhanced by increased frequency and intensity of frontal storms in winter. Porter (1979) felt it likely that the glacier developed from a combination of increased winter snowfall, reduced ablation rates due to lower air temperatures, and reduced solar radiation due to increased cloudiness – a view which contrasts substantially with Selling's (1948) interpretation of lowered TWI and drier high-elevation sites during cool periods.

10.2.4 Galápagos Islands

A pollen and spore record from 16 m of sediments spanning 48 000 years from El Junco lake, at an elevation of 650 m on Isla San Cristóbal (1 °S), gives a good record of paleoclimate for the Galápagos Archipelago (Colinvaux 1972, 1984; Colinvaux and Schofield 1976a,b). There is no evidence of a climate wetter than the present one during the 48 000 years of record. An equivalent of the present dry climate prevailed from 3000 years B.P. to the present and from about 8000 to 6200 B.P.. The intervening period of 3200 years (6200 B.P. to 3000 B.P., a period which correlates partially with the so-called post-glacial climatic optimum) was somewhat drier than now. From 34 000 B.P. to 10 000 B.P., during the time of worldwide cooling, the climate was considerably drier than now, with very little precipitation. [Adsersen (pers. comm.) has raised the point that El Junco lake is very near the level of the inversion layer and could have been in the very dry air above the inversion layer during full glacial conditions when sea level was ca. 60 m lower than today]. The implication is that the stable inversion layer normally present in the current Galápagos environment has been persistent for at least 48 000 years; this inversion layer largely prevents rainfall and creates a fog zone. Heavy rainfall occurs in the Galápagos only during years when strong El Niño/Southern Oscillation (ENSO) conditions prevail. Colinvaux (1972) suggested that the extremely dry conditions in the Galápagos during full glacial indicate an ITCZ consistently north of the equator. This implies a lower incidence of El Niño/ Southern Oscillation (ENSO) conditions and *might* imply a higher frequency of such conditions with oceanic warming.

10.2.5 Easter Island and Other Sites in the Southern Pacific

Flenley et al. (1991) have investigated the Late Quaternary vegetation and climatic history of Easter Island (27 °S). The pollen record goes back to 38 000 years B.P. Numerous plant taxa dominant in the pollen record are now extinct, having been apparently eliminated by the descendents of human colonizers which arrived about 1500 years ago. Particularly noteworthy components of the original vegetation were a species of palm (closely related to a Chilean endemic) and at least six different types of probably woody Asteraceae, perhaps similar to the woody Asteraceae of present-day Hawaii (*Dubautia-Argyroxiphium*) and the Juan Fernandez Islands.

Pollen diagrams from sites on Easter Island (Flenley et al. 1991) provisionally indicate a climate cooler and drier than present in the period between 26 000 and 12 000 years B.P.; a 350-m downward shift of vegetation is suggested, indicating a temperature drop of 1.8–2.8 °C, consistent with a 2 °C cooling for that part of the Pacific estimated by the climate model of Gates (1976). In general, most sites in the southern Pacific region, west to Australia and New Guinea, and including the Cook Islands, Fiji, and New Zealand, appear to have had cooler and drier conditions in Late Pleistocene (Flenley et al. 1991). Temperature depression at the end of last full glacial (18 000–14 000 B.P.) for the North Island of New Zealand (35–40 ° S) is estimated at 4 °C by Newnham et al. (1989). For the last full glacial cycle in Australia, the majority of studies indicate that cool phases have been dry and warm phases wet, but recent dating refinements suggest that temperature and effective precipitation have not been entirely in phase (Kershaw and Nanson 1993).

10.3 Biological Effects of Warming on Islands

In view of the difficulties in predicting the magnitude of climate change in the near future, projections of biological effects are largely premature. However, at least one documented instance of recent warming in the southern hemisphere gives a premonition of how climate change is often likely to favor invasive species, leading to erosion of native biological diversity. On subantarctic Marion Island (47 °S), mean annual temperature has risen by 0.93 °C, from about 5.5 to 6.5 °C over the period 1951–1988 (Smith and Steenkamp 1990). The house mouse (*Mus musculus*) population on the island is strongly temperature-limited and was thought to be near its ecological tolerance limits in terms of temperature prior to the recent warming. There are strong indications that the mouse population has increased substantially in conjuction with the warming and that its predation on soil invertebrates (83–94% of the mouse diet) is increasing dramatically (Smith and Steenkamp 1990).

A similar effect might be expected with warming in the Hawaiian islands, involving the spread of an alien mosquito, *Culex quinquefasciatus*, to higher elevations. The mosquito, a vector of avian malaria and pox believed to contribute in a major way to the decline of Hawaiian honeycreepers (Aves: Fringillidae: Drepanidinae), is currently limited mostly to below 1500 m elevation by cool temperatures. Several species of honeycreepers survive only above this elevation on Maui and Hawaii islands (Scott et al. 1986).

It is highly possible that changes in frequency and intensity of extreme events due to warming climate may have more drastic effects on island ecosystems than changes in average climate. For example, incidence and intensity of hurricanes may increase if tropical sea surface temperature increases (Emanuel 1987; Wendland 1977), resulting in drastic changes in forest canopy structure (O'Brien et al. 1992) and damage to coral ecosystems (Moran and Reaka-Kudla 1991). Canopy opening due to severe windstorms will tend to favor the spread of invasive introduced species. Increase in disturbance by hurricanes could lead to greater disruption of island forests by fire. In conjuction with drought, hurricane-damaged forests can be highly susceptible to fire (e.g., Whigham et al. 1991). If drought conditions were to become more frequent and intense in archipelagoes such as Hawaii, dramatic negative effects from fire are likely to result. In oceanic island situations, drought and fire generally promote the spread of invasive introduced species at the expense of native species (e.g., Atkinson and Greenwood 1972; Smith and Tunison 1992).

10.4 Conclusions: Prognosis for Potential Effects of Global Warming on Island Biological Diversity

There is much disagreement over the amount of sea-surface temperature change in tropical seas over the past 20 000 + years. The view that island climate change will be strongly buffered by the tropical oceans is at best questionable. Research during the next decade will very likely clarify this situation. Meanwhile, the following generalizations are possible:

1. Species temperature tolerance zones (vegetation zones?) for both native and introduced species may be shifted upward by roughly 200–400 m, assuming an increase in mean air temperature at sea level of 1–2 °C and a lapse rate of 0.5 °C/100 m rise in elevation. This shift is likely to be much larger if tropical seas undergo more drastic temperature fluctuations as global climate changes.
2. Mean precipitation may increase, decrease, or stay the same, depending on the island. We can speculate that mean precipitation may tend to increase in the Galápagos, as a result of warming in the Pacific and possible weakening of the cold Humboldt Current. If this means a higher frequency of El Niño/Southern Oscillation conditions with oceanic warming, more frequent drought in Hawaii might be expected (Chu 1989) as well as other dramatic influences on climate worldwide. In the Hawaiian Islands, complex changes in precipitation patterns may be expected with increases in the mean values for some sites and decreases for others; because of steep climatic gradients, the change for any particular location may be substantial. A particularly dynamic zone may be the level of the trade-wind inversion. Clear evidence of Late

Quaternary glaciation on Mauna Kea shows that unexpectedly large climatic fluctuations have occurred at upper elevations in the Hawaiian Islands in the recent past and may occur in the future.

3. Changes in frequency and intensity of extreme events, such as hurricanes or drought, due to warming climate, are likely to have more drastic effects on island ecosystems than changes in average climate.
4. The above-mentioned changes will generally favor invasive introduced species in competition with native/endemic island species. The greater the magnitude of the changes, the more the invasive species will be favored.

Acknowledgments. I appreciate suggestions by Lynn Kutner, Hall Cushman, and Thomas Giambelluca of references useful for this review. Peter Vitousek made constructive suggestions for improvement of the manuscript.

References

Aharon P (1983) 140,000-yr isotope climatic record from raised coral reefs in New Guinea. Nature 304: 720–723
Atkinson IAE, Greenwood RM (1972) Effects of the 1969–70 drought on two remnants of indigenous lowland forest in the Manawatu district. Proc N Z Ecol Sci 19: 34–45
Blumenstock DI, Price S (1967) Climates of the States: Hawaii. Climatography of the United States, No 60–51, US Department of Commerce Washington, DC, 27 pp
Brockie R, Loope LL, Usher MB, Hamann O (1988) Biological invasions of island nature reserves. Biol Conserv 44: 9–36
Chu P-S (1989) Hawaiian drought and the southern oscillation. Int J Climatol 9: 619–631
Chuey JM, Rea DK, Pisias NG (1987) Late Pleistocene paleo-climatology of the Central Equatorial Pacific: a quantitative record of eolian and carbonate deposition. Quat Res 28: 323–339
CLIMAP Project Members (1976) The surface of the ice-age earth. Science 191: 1131–1137
Colinvaux PA (1972) Climate and the Galápagos Islands. Nature 240: 17–20
Colinvaux PA (1984) The Galápagos climate: present and past. In: Perry R (ed) Galápagos. Pergamon Press, Oxford, pp 55–69
Colinvaux PA (1993) Ecology 2. John Wiley & Sons, New York, 688 pp
Colinvaux PA, Schofield EK (1976a) Historical ecology in the Galápagos Islands. I. A Holocene pollen record from El Junco lake, Isla San Cristóbal. J. Ecol 64:989–1012
Colianvaux PA, Schofield EK (1976b) Historical ecology in the Galápagos Islands. II. A Holocene spore record from El Junco lake, Isla San Cristóbal. J. Ecol 64: 1013–1028
Dorn RI, Phillips FM, Zreda MG, Wolfe EW, Jull AJT, Donohue DJ, Kubik PW, Sharma P (1991) Glacial chronology. Natl Geogr Res Expl 7(4): 456–471
Emanuel KA (1987) The dependence of hurricane intensity on climate. Nature 326: 483–485
Emiliani C (1992) Pleistocene paleotemperatures. Science 257: 1462
Flenley JR, King ASM, Jackson J, Chew C, Teller JT, Prentice ME (1991) The Late Quaternary vegetational and climatic history of Easter Island. J Quat Sci 6(2): 85–115
Gates WL (1976) Modelling the ice-age climate. Science 191: 1138–1144
Gavenda RT (1992) Hawaiian Quaternary paleoenvironments: a review of geological, pedological, and botanical evidence. Pac Sci 46: 295–307
Giambelluca TW, Nullet D (1991) Influence of the trade-wind inversion on the climate of a leeward mountain slope in Hawaii. Climate Res 1: 207–216
Giambelluca TW, Nullet M, Schroeder T (1986) Rainfall atlas of Hawaii. Report 76, Hawaii Dept of Land and Natural Resources, Div of Water and Land Development Honolulu

Kershaw AP, Nanson GC (1993) The last full glacial cycle in the Australian region. Global Planet Change 7: 1–9
Loope LL, Mueller-Dombois D (1989) Characteristics of invaded islands, with special reference to Hawaii. In: Drake JA, Mooney HA, di Castri F, Groves RH, Kruger FJ, Rejmanek M, Williamson M (eds) Biological invasions: a global synthesis. SCOPE 37. John Wiley Chichester, pp 257–280
Lyons SW (1982) Empirical orthogonal function analysis of Hawaiian rainfall. J Appl Meteorol 21: 1713–1729
Manabe S, Hahn DG (1977) Simulation of the tropical climate of an ice age. J Geophys Res 82: 3889–3911
Moran DP, Reaka-Kudla ML (1991) Effects of disturbance and enhancement of coral reef cryptofaunal populations by hurricanes. Coral Reefs 9: 215–224
Newnham RM, Lowe DJ, Green JD (1989) Palynology, vegetation and climate of the Waikato lowlands, North Island, New Zealand, since c. 18,000 years ago. J R Soc NZ 19(2): 127–150
O'Brien ST, Hayden BP, Shugart HH (1992) Global climate change, hurricanes, and a tropical forest. Climat Change 22: 175–190
Peters RL, Darling JDS (1985) The greenhouse effect and nature reserves. BioScience 35: 707–717
Porter SC (1979) Hawaiian glacial ages. Quat Res 12: 161–187
Potter GL, Ellsaesser HW, MacCracken MC, Luther FM (1975) Possible climatic impact of tropical deforestation. Nature 258: 697–698
Price S (1983) Climate. In: Armstrong RW (ed) Atlas of Hawaii, 2nd edn. University of Hawaii Press, Honolulu, pp 59–66
Rasmussen EM, Wallace JM (1983) Meteorological aspects of the El Niño/Southern Oscillation. Science 222: 1195–1202
Rind D, Peteet D (1985) Terrestrial conditions at the last glacial maximum and CLIMAP sea-surface temperature estimates: are they consistent? Quat Res 24: 1–22
Schneider SH, Mearns L, Gleick PH (1992) Climate-change scenarios for impact assessment. In: Peters RL, Lovejoy TE (eds) Global warming and biological diversity. Yale University Press, New Haven pp 18–55
Scott JM, Mountainspring S, Ramsey FL, Kepler CB (1986) Forest bird communities of the Hawaiian Islands: their dynamics, ecology, and conservation. Cooper Ornithological Society, Studies in Avian Biology No 9. Allen Press, Lawrence, Kansas
Selling OH (1948) Studies in Hawaiian pollen statistics. Part III. On the Late Quaternary history of the Hawaiian vegetation. Bernice P Bishop Mus Spec Publ 39: 154 pp + 27 plates
Servant M, Maley J, Turcq B, Absy M-L, Brenac P, Fournier M, Ledru M-P (1993) Tropical forest changes during the Late Quaternary in African and South American lowlands. Global Planet Change 7: 25–40
Smith CW, Tunison JT (1992) Fire and alien plants in Hawaii: research and management implications for native ecosystems. In: Stone CP, Smith CW, Tunison JT (eds) Alien plant invasions in native ecosystems of Hawaii: management and research. Univ Hawaii Coop Natl Park Studies Unit, Honolulu, pp 394–408
Smith SV, Buddemeir RW (1992) Global change and coral reef ecosystems. Annu Rev Ecol Syst 23: 89–118
Smith VR, Steenkamp M (1990) Climatic change and its ecological implications at a subantarctic island. Oecologia 85: 14–24
Stute M, Schlosser P, Clark JF, Broecker WS (1992) Paleotemperatures in the southwestern United States derived from noble gases in ground water. Science 256: 1000–1003
Webb T (1992) Past changes in vegetation and climate: lessons for the future. In: Peters RL, Lovejoy TE (eds) Global warming and biological diversity. Yale University Press, New Haven pp 59–75
Wendland WM (1977) Tropical storm frequencies related to sea surface temperature. J Appl Meteorol 16: 477–486
Whigam DF, Olmsted I, Cano EC, Harmon ME (1991) The impact of Hurricane Gilbert on trees, litterfall, and woody debris in a dry tropical forest in the northeastern Yucatan Peninsula. Biotropica 23: 434–441

Section C
Diversity and Ecosystem Function

11 Ecosystem-Level Consequences of Species Additions and Deletions on Islands

J. HALL CUSHMAN

11.1 Introduction

Human activities are changing natural communities and ecosystems at unprecedented rates and spatial scales. Two of the most alarming of these modifications are the exchange of biota among previously isolated regions (introductions) and the loss of native populations and entire species (extinctions). The abundance of species additions and deletions has led to growing concern about the relationship between biological diversity and the capacity of ecosystems to function and to provide the services on which humans depend (Ehrlich and Mooney 1983; Ehrlich and Wilson 1991; Kareiva et al. 1993; Lawton and Jones 1993; Schulze and Mooney 1993).

For well over a century, ecologists have recognized that species additions and deletions are a fundamental feature of natural communities and ecosystems. For example, the entire concept of succession is based on this premise (McIntosh 1981) as is the equilibrium theory of island biogeography (MacArthur and Wilson 1963, 1967). However, humans have dramatically accelerated the rates at which additions and deletions occur (Diamond and Case 1986; Drake et al. 1989; D'Antonio and Dudley, Chap. 9 this Vol.). Estimating these rates is exceedingly difficult, primarily because most introductions and extinctions have not been well documented. For example, Smith et al. (1993) and others have begun to compile global data sets on the number of human-caused additions and deletions, but their numbers are most certainly underestimates.

The ecosystem-level consequences of these biotic changes will be determined by the capacity of individual species to influence ecosystem processes. Hence, much of this chapter will be devoted to discussing the degree to which particular taxa can be classified as keystone species – species that have disproportionately large influences on the composition and persistence of biotic assemblages – and whether or not these effects scale up to the ecosystem level. Paine (1966, 1969) first proposed the keystone species concept, emphasizing the importance of predators as mediators of the composition and stability of intertidal communities. Other ecologists have subsequently broadened the concept to include influential herbivores, pathogens, competitors, and mutualists (Gilbert 1980; Bond 1993). Some species undoubtedly have disproportionately large influences on the ecosystems

Department of Biology, Sonoma State University, Rohnert Park, California 94928, USA

Ecological Studies, Vol. 115
Vitousek et al. (eds) Islands

they inhabit (see reviews by Vitousek 1986; Naiman 1988; Hobbie 1992; Bond 1993), while others have only minimal influences, or their influences appear to be equivalent to those of others species. This dichotomy led Walker (1992) and Lawton and Brown (1993) to develop the "redundancy hypothesis" which proposes that biological systems contain only a few species that influence ecosystem-level processes (the "drivers" according to Walker 1992), with the remaining majority having little or no effects at the ecosystem level (the "passengers").

The majority of human-caused species additions and deletions have not been well documented, either in terms of their numbers or effects (see Diamond and Case 1986; Drake et al. 1989; Smith et al. 1993). As a result, it is usually difficult to determine rigorously the ecosystem consequences of these changes, let alone those for populations and communities. Alternative sources of insight can be provided by basic ecological studies that have assessed the short- and long-term effects of adding and/or removing particular species. Unfortunately, such studies can have their limitations, as they are usually conducted by population ecologists whose primary emphasis is not the functional role of species in ecosystems (Vitousek 1986, 1990; Lawton and Brown 1993; Woodward 1993).

In this chapter, I have five objectives. First, I discuss how our understanding of the functional role of species has been hindered by differences in the ways that population and ecosystem ecologists study nature. Second, I briefly review our knowledge of influential species on islands – both native and exotic – as a means of assessing the consequences of both potential and actual species additions and deletions. Third, I propose a set of probabilistic rules – and discuss their limitations – that may be of use in predicting the ecosystem consequences of species additions and deletions. Fourth, I propose and discuss two predictions regarding the effects of additions and deletions on islands versus the mainland. Fifth, I conclude by emphasizing the importance of variation in the effects of individual species on ecosystem processes and the need for caution when categorizing species with respect to their functional importance.

11.2 Linking Biodiversity to Ecosystem Processes

Although there are many noteworthy exceptions, researchers in population and ecosystem ecology have traditionally studied nature in fundamentally different ways. In the words of Lawton and Jones (1993), the two groups"... have ploughed their own independent furrows and developed their own paradigms, approaches and questions." These differences may go a long way toward explaining our generally poor understanding of the influence of biological diversity on ecosystem-level processes (see discussion of this topic by Vitousek 1986, 1990; Lawton and Brown 1993; Lawton and Jones 1993; Vitousek and Hooper 1993). Historically, population ecologists have not often considered the influences of biodiversity on ecosystem processes, while ecosystem ecologists infrequently considered the effects of individual taxa on ecosystem processes. Thus, one group studied the "desired" biotic units (species, assemblages) but not the desired response variables (ecosystem processes), while the other studied the desired response variables but considered biotic units at too coarse a level (forests, landscapes).

The work of Brown and Heske (1990) clearly illustrates this situation. In a 12-year experimental study, they showed that the removal of three ecologically similar species of kangaroo rat (*Dipodomys*) caused habitats to shift from desert shrub to arid grassland. Almost certainly, these impressive effects were accompanied by a host of biogeochemical and hydrological effects, but the authors did not quantify them (as they were interested in community-level questions). However, the authors did note increased litter accumulation and snowfall persistence on removal plots. Thus, in this system, we hypothesize that kangaroo rats have a strong influence on ecosystem properties – and their permanent deletion from this ecosystem would have large consequences – but we lack ecosystem-level data to back up this claim.

11.3 Additions and Deletions on Islands

As discussed initially by MacArthur and Wilson (1967), a fundamental feature of island assemblages is that they are frequently characterized by high rates of species turnover, due to patterns of immigration (additions) and extinctions (deletions) that are mediated by island size and isolation. With human settlement, there has been a dramatic acceleration in the rates of species additions and deletions. While numerous studies have shown that these changes can have major effects on island communities (Diamond and Case 1986; Mooney and Drake 1986; Drake et al. 1989; D'Antonio and Dudley, Chap. 9 this Vol.), most do not provide data on ecosystem-level processes (but see Vivrette and Muller 1977; Vitousek et al. 1987; D'Antonio and Vitousek 1992; D'Antonio and Dudley, Chap. 9 this Vol.). For example, the avifauna of Guam has been decimated, with 17 of its 18 native bird species now rare or presumed extinct. Savige (1987) has provided convincing evidence suggesting that the introduction of an avian predator, the brown tree snake (*Boiga irregularis*), was largely responsible for these dramatic effects. The snake now appears to be having a similar effect on Guam's nocturnal lizards (Rodda and Fritts 1992). However, despite the magnitude of these effects, it remains unclear, and largely unexplored, if and how these changes in vertebrate biodiversity translate into ecosystem-level effects.

11.3.1 Functional Properties of Island Species

Below, I discuss three of the best-documented examples of the functional properties of species in island ecosystems. Before doing so, I should mention that Vitousek and colleagues (Vitousek et al. 1987; Vitousek and Walker 1989;

Vitousek 1990) have presented some of the best data to date on the effects of a species addition. Because their work on the nitrogen-fixing tree *Myrica faya* in Hawaii has been reviewed elsewhere, I will not discuss it in this section. However, I will touch on this system in later sections that attempt to predict the ecosystem-level effects of species additions and deletions.

11.3.1.1 Land Crabs and Snails on Christmas Island

Land crabs can have dramatic influences on the ecosystem processes of oceanic islands. The most compelling evidence for this comes from work completed on Christmas Island (360 km south of Java) with the native red land crab, *Gecarcoidea natalis*. The abundance of these large crabs is extraordinary, with densities ranging from 0.8–2.6 m^{-2} and biomass averaging 0.8 metric tons ha^{-1} (O'Dowd and Lake 1989). They are dominant consumers on the rain forest floor, with a broad diet that includes seeds, fruits, seedlings, and leaf litter. Red crabs are important herbivores of tree and vine seedlings, defoliating 29–35% of all uncaged seedlings of 18 species in just a few days (O'Dowd and Lake 1990). They also rapidly remove large numbers of fruits and seeds from the forest floor and relocate them to their burrows (O'Dowd and Lake 1991). For example, field experiments showed that crabs removed large proportions of plant propagules for all 17 species tested, usually within 12 h of placement. Through eating and returning leaves to burrows, red crabs also removed 30–50% of the leaf fall in the rainforest (O'Dowd and Lake 1989). More recently, the longer-term experiments of Green (1993) showed that red crabs remove 39–86% of the annual leaf litter. Presumably because of this, the soil near burrow entrances is characterized by significantly higher concentrations of organic matter and a range of mineral nutrients (N, P, K, Ca, Na, and Mg).

In addition to having pronounced effects on island vegetation, red crabs reduce or prevent the invasion (i.e., addition) of the African land snail, *Achatina fulica* (Lake and O'Dowd 1991). These snails cause tremendous problems on many tropical islands (Mead 1992) and may significantly alter such ecosystems. Lake and O'Dowd (1991) clearly showed that red crabs and African land snails exhibit an inverse pattern of distribution on Christmas Island, with snails being rare in the crab-dominated rainforest and common in the less populated disturbed areas that border the forest. Transplant experiments strongly supported the hypothesis that this pattern was due to far greater crab predation on snails in the forest than in bordering areas.

This work strongly suggests that the deletion of land crabs from Christmas Island, or the addition of red crabs to islands where they are not currently found, would have dramatic effects on ecosystem processes. While the red crab has a limited distribution, there is evidence to suggest that other land and mangrove crabs are influential components of many tropical oceanic islands (Alexander 1979; Louda and Zedler 1985; Robertson 1991).

11.3.1.2 Manuring, Moths, and Mice on Marion Island

Ecosystem processes on subantarctic islands have received considerable attention during the past 25 years, due largely to the activities of the Scientific Committee on Antarctic Research (see Siegfried et al. 1985; Kerry and Hempel 1990). Marion Island, located 1600 km south of the southern tip of Africa, has been particularly well studied and provides some of the best data available on the influence of island biodiversity on ecosystem processes. Four groups of organisms play particularly key roles in this ecosystem: sea birds, seals, flightless moths, and introduced mice.

A variety of birds (Wandering Albatross, Giant and Burrowing Petrel, and King and Gentoo Penguin) and mammals (Elephant Seal) feed in the ocean surrounding Marion Island and deposit large amounts of guano, dung, and urine on its terrestrial habitats. Such manuring has strong influences on the availability of nitrogen and phosphorous in the soil (see Smith 1978). For example, 5 months after Wandering Albatrosses have established a nest, there is a fivefold increase in the organic nitrogen content of the surrounding soil. Associated with this influx of nutrients are substantial increases in percent cover and primary production, as well as changes in the species composition of plant communities. For islands in general, I suggest that manuring organisms are likely to exert large influences on ecosystem processes, given the high levels of abundance they commonly attain and the high edge-to-area ratios that characterize islands.

Invertebrate detritivores are especially important in ecosystems like those found on Marion Island, where virtually all primary production becomes dead organic matter, due to the absence of vertebrates herbivores and the insignificance of insect herbivory (Smith 1978; Smith and Steenkamp 1990). Larvae of the endemic flightless moth *Pringleophaga marioni* are the most abundant litter-dwelling macroinvertebrates on the island (mean biomass is 9.3 kg ha^{-1} $year^{-1}$), although slugs, weevil larvae, earthworms, and snails are also important detritivores. On a vegetated coastal plain, Crafford (1990) estimated that *P. marioni* processes at least 1500 kg ha^{-1} $year^{-1}$ of leaf litter, which is 50% of the maximum primary production reported for the island. Such processing by moth larvae considerably enhances microbial activity and subsequent release of essential nutrients. For example, Smith and Steenkamp (1990) performed experiments showing that the addition of just two moth larvae into litter-containing microcosms stimulated mineralization of nitrogen tenfold and phosphoros threefold. Crafford (1990) has also performed such experiments with similar results. Thus, the primary mediators of nutrient mineralization on this island are a range of soil macroinvertebrates, with *P. marioni* being the most abundant. The deletion of such a keystone species could have dramatic effects on the ecosystem processes on this island.

Introduced by sealers in 1818, house mice (*Mus musculus*) have altered decomposition and nutrient cycling on Marion Island (see Crafford 1990; Smith and Steenkamp 1990). The mice prey on a range of detritus-feeding invertebrates, and *P. marioni* comprises 50–75% of their diet throughout the year. Through their predation on *P. marioni*, Crafford (1990) estimated that introduced mice prevented the decomposition of at least 1000 kg ha^{-1} $year^{-1}$ of plant litter. Thus, in the absence of mice, litter processing by the moth alone would be 2500 kg ha^{-1} $year^{-1}$. Clearly, the addition of mice has strongly altered ecosystem processes on Marion Island. The importance of predation is further supported by the dramatic differences in population size and composition of the insect detritivore fauna between Marion Island and neighboring Prince Edward Island, which lacks mice (see Crafford and Scholtz 1987).

11.3.1.3 Moose on Isle Royale

Moose, *Alces alces*, have profound influences on the ecosystem processes of Isle Royale (Michigan, USA), an island in Lake Superior that is characterized predominately by boreal forest vegetation (see Pastor et al. 1988, 1993; McInnes et al. 1992). Moose preferentially browse on the growing shoots of deciduous shrubs and trees, and a single adult consumes from 3000–5000 kg of dry matter per year. Documenting the pivotal role played by moose on Isle Royale has been possible largely because of experimental exclosures that have been in place since 1948. This experimental program has revealed that moose browsing substantially alters forest structure, species composition, nutrient cycling, net primary production, and successional pathways.

Moose stimulate a cascading series of events in this island ecosystem (Pastor et al. 1993). They feed preferentially on early successional tree species, such as aspen, balsam poplar, and paper birch, while avoiding those of later successional stages, such as spruce and balsam fir (Pastor et al. 1988). Such browsing results in forests that have significantly fewer canopy trees and a more developed understory of shrubs and herbs, compared to forests where moose have been excluded (McInnes et al. 1992). Over time, browsing causes shifts in species composition, with forests becoming increasingly dominated by spruce (Krefting 1974). The transition from hardwood to spruce significantly reduces both the quantity and quality of litter reaching the forest floor: slow-growing spruce have higher rates of leaf retention, slower rates of leaf decomposition, and lower nutrient content in leaves. As a result, nutrient availability (including rates of nitrogen mineralization) and microbial activity are significantly reduced in browsed plots, and these changes in soil chracteristics subsequently reduce net primary production (see Pastor et al. 1993). Thus, moose have major influences on the ecosystem processes of Isle Royale, and the many noninsular areas where they also occur (see Molvar et al. 1993).

11.4 Predicting Consequences of Additions and Deletions

11.4.1 Probabilistic Rules

Given the high incidence of species additions and deletions, particularly on islands (see MacDonald et al. 1989; Smith et al. 1993), it is imperative to understand the ecosystem consequences of human-caused alterations in biological diversity. Ideally, one would like to formulate a set of probabilistic rules that could be used to predict whether or not species have ecosystem-level effects. Below, I propose two possibilities and then discuss some of their potential shortcomings. Both possibilites concern functional groups, which are defined as groups of species that influence an ecosystem process in similar ways (Schulze and Mooney 1993). In most cases, a given species will be a member of several functional groups, as it usually affects more than one ecosystem process. With this concept, it is useful to consider the diversity of functional groups present in an ecosystem, the abundance of species within each functional group (which I refer as "depth"), and the influence of human activities on both the diversity and depth of these groups. As discussed latter in this paper, I view "depth" – rather than "redundancy" (sensu Walker 1992; Lawton and Brown 1993) – as a far more useful and accurate shorthand for the abundance of species within functional groups.

Rule 1. In order to have ecosystem-level effects, a species must be abundant relative to other taxa in its functional group and exhibit an ecosystem-wide distribution. Although Rule 1 is fairly straightforward, it is nevertheless essential: without significant abundance and distribution, it is highly unlikely that a species will have large ecosystem effects, even if it has strong local effects where it occurs. As outlined above, land crabs, manuring vertebrates, flightless moths, and house mice are all taxa with ecosystem-level effects that follow the predictions of Rule 1. However, simply being abundant and widespread does not insure that a species will have large ecosystem effects, i.e., Rule 1 is necessary but not sufficient.

Rule 2. In order to have ecosystem-level effects, a species must either (1) directly play a unique or under-represented functional role or (2) indirectly play such a role by significantly altering the abundance and/or distribution (and therefore influence) of other taxa that play unique or under-represented functional roles. Here, the emphasis is on the need to simultaneously consider the functional roles of species combined with ecosystem characteristics, such as the abundance of species within a particular functional group (i.e., the degree of "depth"). Moths, red crabs, and moose each provide clear examples of taxa that directly play underrepresented functional roles in the ecosystems they inhabit. By altering the abundance of moths on Marion Island, house mice are an example of the indirect pathway. Some species may play important functional roles through both pathways. For example, in Hawaii, the invasive nitrogen-fixing tree *Myrica faya* plays a direct functional role by quadrupling nitrogen availability in nutrient-poor volcanic soils and may play an indirect functional role by permitting subsequent invasions by exotic plant species that require more fertile soils (Vitousek et al. 1987; Vitousek and Walker 1989).

11.4.2 Problems with Probabilistic Rules

Many factors may interfere with attempts to generate probabilistic rules for predicting whether or not particular species significantly affect ecosystem-level processes. For example, while abundance and widespread occurrence may often be a prerequisite for species with ecosystem-level effects (Rule 1), some organisms – particularly vertebrates – may commonly prove to be exceptions. For some species, single individuals or small groups can cause large-scale changes to landscapes. Beaver (*Caster canadensis*) are a classic example: just a handful of individuals inhabiting a riparian system can modify the size and characteristics of wetlands, nutrient cycling, and decomposition, and the species composition of entire communities – often leading to physical and biotic changes that persist for centuries (Naiman et al. 1988). In addition, a range of other vertebrates – from pocket gophers to moose – may regularly exert large influences on ecosystems without being particularly abundant or widespread.

Compensatory responses may be another factor that interferes with attempts to accurately predict the ecosystem-level consequences of species additions and deletions. The abundance or size of other species in the ecosystem may increase (to take advantage of increased resource availability) or decrease in response to species additions and deletions. For example, what would be the ecosystem-level consequences of deleting moths (*P. marioni*) from Marion Island? Based on Rules 1 and 2, I would predict a considerable decrease in rates of decomposition and thus the availability of nitrogen and phosphorus. However, if previously rare but functionally similar detritivores – such as slugs whose activity on a per capita basis leads to much greater nutrient release (Smith and Steenkamp 1990) – responded to the deletion of moth larvae by significantly increasing in abundance and/or distribution, then there might be little or no ecosystem-level effect.

Spatial and temporal variation in the effect of species on ecosystem processes is probably the most important factor that will interfere with attempts to generate predictive rules. It is unrealistic to assume that the influences of individual species on ecosystems will remain constant in space and time. Biotic and abiotic settings have tremendous influences on the outcome of species interactions (e.g., Thompson 1988; Cushman and Whitham 1989; Cushman and Addicott 1991), and on the effects of particular species on entire communities and ecosystems (see Mooney and Drake 1986; Drake et al. 1989). For example, the effects of moose on Isle Royale ecosystems varies considerably between sites on the island (Pastor et al. 1988). This variation is due in part to among-site differences in moose densities, and to among-site differences in the initial species composition of the forest (caused by different fire regimes). Clearly then, there is a "distribution of outcomes" (sensu Thompson 1988) with respect to the effects of species additions and deletions on ecosystem processes.

11.5 Island-Mainland Comparisons

Island biota differ from those on the mainland in many ways, and two of these differences are particularly relevant to a discussion of the ecosystem-level effects of species additions and deletions. First, because of their reduced size and increased isolation, islands commonly have fewer species than mainland areas of comparable size (MacArthur and Wilson 1963, 1967). Second, largely due to isolation and the variable dispersal abilities of potential colonists, islands often lack the full compliment of taxonomic groups found in comparable noninsular regions (Carlquist 1965; Eliasson, Chap. 4, Vol.). This latter characteristic has been referred to taxonomic "disharmony" and a widely cited example of it is the absence of ants in the Hawaiian Islands (e.g., Cole et al. 1992).

As a result of these two well-known differences, islands should have fewer species per functional group and fewer functional groups than equivalent mainland areas. Based on these generalities, I offer two predictions concerning the responses of assemblages – and the subsequent ecosystem-level effects – following species deletions and additions.

Prediction 1. Following a given species deletion, substitution by functionally equivalent species will be less likely on islands than comparable mainland because there should be fewer species per functional group on islands. The consequences of this situation will depend on whether the deleted species, for which there is a smaller pool of potential substitutes, was abundant, widespread in occurence, and played a unique or under-represented functional role in the ecosystem (i.e., Rules 1 and 2). Strong evidence to support or refute this prediction are not available, but flying foxes on many Pacific islands may offer preliminary insight. Cox et al. (1991) have hypothesized that flying foxes are keystone mutualists (sensu Gilbert 1980), through their effect on the reproductive success of a large proportion of the plant species on these islands. The authors cite the flying fox *Pteropus samoensis*, which on the island of Samoa appears to be the single most important seed disperser, generating 80–100% of the dry-season seed rain, and is also thought to be the most influential pollinator. Flying foxes are being eliminated from Pacific islands, largely due to the combined effects of habitat loss, over-hunting, and introduced predators. The ecosystem-level effects of this reduction and probable deletion, if any, depend on (1) the availability of other mutualists that can substitute for the beneficial services provided by flying foxes, and (2) the importance of these services – pollination and seed dispersal – for ecosystem-level processes. For example, does the reduction or elimination of flying foxes substantially alter the composition of dominant plant species? If so, one would expect to detect changes in various biogeochemical processes, similar to those shown for moose on Isle Royale (Pastor et al. 1988). However, it is worth mentioning that, to my knowledge, there are no studies that have shown that changes in the abundance and composition of pollinators or seed dispersers have led to changes in the composition of plant communities. This lack of evidence may occur not because pollinators and seed dispersers do not have effects at the ecosystem level, but because (1) previous investigators have failed to consider this possibility or (2) the temporal scales of most studies are usually insufficient to detect changes in recruitment and species composition.

Prediction 2. Species introduced to islands will be more likely to play unique or underrepresented functional roles than on the mainland. Because islands often lack a full compliment of functional groups, there is greater probability that species added to islands by humans will be members of "missing" or under-represented functional groups. For example, prior to 1800, early successional plant communities on nitrogen-deficient volcanic soils in Hawaii lacked dominant, nitrogen-fixing plants. Since then, the nitrogen-fixing tree *Myrica faya* has readily invaded the region, often forming monospecific stands. By greatly increasing the availability of nitrogen in the soil, *Myrica* has influenced productivity, nutrient cycling, and perhaps the rate and direction of primary succession (Vitousek et al. 1987; Vitousek and Walker 1989). Thus, the absence or underrepresentation of the nitrogen-fixing functional group from these early successional communities (presumably due to taxonomic "disharmony") appears to explain why *Myrica* has played such a dominant functional role in this island ecosystem.

11.6 Functional "Redundancy"

There has been a flurry of discussion about functional "redundancy" in ecosystems (e.g., Lawton and Jones 1993) since this idea was proposed by Walker (1992) and further developed by Lawton and Brown (1993). Given the apparent popularity of this topic, I will conclude by discussing two issues that are especially relevant to this discussion. First, central to the idea of functional "redundancy" is the degree of "depth" within functional groups. As outlined earlier, I use depth as shorthand for the abundance of species within each functional group. Such depth is what some refer to as "redundant" biodiversity. However, this view fails to consider the importance of depth within functional groups following natural and human-caused disturbances. Rather than being considered redundant, depth within functional groups may turn out to be a vital property of ecosystems because it allows systems to better withstand and/or recover from the increasing perturbations (the "insurance" hypothesis). In addition, depth within functional groups may prove to be especially important on islands (and insular environments in general) where recolonization following disturbance is restricted by isolation.

The second issue relating to "redundancy" deals with the complexity of ecosystems and our zeal to assign functional properties to individual species. Use of the term redundancy implies that we can accurately categorize taxa with respect to their ecosystem-level effects. However, in this chapter, I have discussed various factors that may interfere with such attempts to categorize species. Following additions and/or deletions, members of a community may compensate in ways that increase, decrease, or eliminate the effects of the initial alterations. In addition, the biotic and abiotic setting in which species are added to and deleted from ecosystems will have considerable influences on the effect of these changes. Furthermore, ascribing functional roles will be made difficult by pronounced variation in the rates at which species or assemblages influence ecosystem processes. Much of the variation can be attributed to whether the species or assemblages in question interact with other species through direct versus indirect pathways. However, there is no need for the rates of expression and magnitude of effects to be correlated. As discussed for the study by Brown and Heske (1990), kangaroo rats exert a tremendous influence on the community, and presumably the ecosystem, but the shift from desert shrub to desert grassland took over a decade to occur. Furthermore, in this same granivore system, Brown et al. (1986) showed that the short-term direct effects were quite different from the longer-term indirect effects. Given these complex and variable ways in which the ecosystem-level effects of species are expressed, ecologists need to exhibit considerable caution when drawing conclusions as to whether or not particular species are "redundant" in terms of their functional roles in ecosystems.

Acknowledgments. I thank Amy Austin, Carla D'Antonio, George Koch, David Hooper, Yiqi Luo, Vanessa Rashbrook, Osvaldo Sala, and Peter Vitousek for comments on this manuscript and/or valuable discussion. I am also indebted to Dennis O'Dowd and Peter Green for providing information on red crabs, and to the Scientific Committee on Problems of the Environment (SCOPE) for financial support during the preparation of this manuscript. Special thanks go to Pericles Maillis and the Bahamas National Trust for their generous hospitality throughout the SCOPE workshop.

References

Alexander HGL (1979) A preliminary assessment of the role of the terrestrial decapod crustaceans in the Aldabran ecosystem. Philos Trans Soc Lond B 286: 241–246

Bond WJ (1993) Keystone species. In: Schulze E-D, Mooney HA (eds) Biodiversity and ecosystem function. Springer, Berlin Heidelberg New York, pp

Brown JH, Heske EJ (1990) Control of a desert-grassland transition by a keystone rodent guild. Science 250: 1705–1707

Brown JH, Davidson DW, Munger JC, Inouye RS (1986) Experimental community ecology: the desert granivore system. In: Diamond J, Case TJ (eds) Community ecology. Harper and Row, New York, pp 41–61

Carlquist S (1965) Island life: a natural history of the islands of the world. Natural History Press, New York

Cole FR, Medeiros AC, Loope LL, Zuehlke WW (1992) Effects of the Argentine ant on arthropod fauna of Hawaiian high-elevation shrubland. Ecology 73: 1313–1322

Cox PA, Elmqvist T, Pierson ED, Rainey WE (1991) Flying foxes as strong interactors in South Pacific island ecosystems: a conservation hypothesis. Conserv Biol 5: 448–454

Crafford JE (1990) The role of feral house mice in ecosystem functioning on Marion Island. In: Kerry KR, Hempel G (eds) Antarctic ecosystems: ecological change and conservation. Springer, Berlin Heidelberg, New York, pp 359–364

Crafford JE, Scholtz CH (1987) Quantitative differences between the insect fauna of sub-antarctic Marion and Prince Edward Islands: a result of human intervention. Biol Conserv 40: 255–262

Cushman JH, Addicott JF (1991) Conditional mutualism in ant-plant-herbivore mutualisms. In: Huxley CR, Cutler DF (eds) Ant-plant interactions. Oxford University Press, Oxford

Cushman JH, Whitham TG (1989) Conditional mutualism in a membracid-ant association: temporal, age-specific, and density-dependent effects. Ecology 70: 1040–1047

Diamond J, Case TJ (1986) Overview: introductions, extinctions, exterminations, and invasions. In: Diamond J, Case TJ (eds) Community ecology. Harper and Row, New York, pp 65–79

Drake JA, Mooney HA, de Castri F, Groves RH, Kruger FJ, Rejmanek M, Williamson M (1989) Biological invasions: a global perspective. Wiley, New York
Ehrlich PR, Mooney HA (1983) Extinction, substitution, and ecosystem services. Bioscience 33: 248–254
Ehrlich PR, Wilson EO (1991) Biodiversity studies: science and policy. Science 253: 758–762
Gilbert LE (1980) Food web organization and conservation of neotropical diversity. In: Soule ME, Wilcox BA (eds) Conservation biology: an evolutionary-ecological perspective. Sinauer, Sunderland, MA, pp 11–34
Green PT (1993) The role of red land crabs (*Gecarcoidea natalis* (Pocock, 1988) Brachyura, Gecarcinidae) in structuring rain forest on Christmas Island, Indian Ocean. PhD Thesis Monash University, Clayton, Victoria, Australia
Hobbie SE (1992) Effects of plant species on nutrient cycling. Trends Ecol Evol 7: 336–339
Kareiva PM, Kingsolver JG, Huey RB (1993) Biotic interactions and global change. Sinauer, Sunderland, MA
Kerry KR, Hempel G (1990) Antarctic ecosystems: ecological change and conservation. Springer, Berlin Heidelberg, New York
Krefting LW (1974) The ecology of the Isle Royale moose with special reference to habitat. Univ Minn Agric Exp Stn Tech Bull 297
Lake PS, O'Dowd DJ (1991) Red crabs in rain forest, Christmas Island: biotic resistance to invasion by an exotic snail. Oikos 62: 25–29
Lawton JH, Brown VK (1993) Redundancy in ecosystems. In: Schulze E-D, Mooney HA (eds) Biodiversity and ecosystem function. Springer, Berlin Heidelberg New York, pp 255–270
Lawton JH, Jones CG (1993) Linking species and ecosystem perspectives. Trends Ecol Evol 8: 311–313
Louda SM, Zedler PH (1985) Predation in insular plant dynamics: an experimental assessment of postdispersal fruit and seed removal, Enewetak Atoll, Marshall Islands. Am J Bot 72: 438–445
MacArthur RH, Wilson EO (1963) An equilibrium theory of insular zoogeography. Evolution 17: 373–387
MacArthur RH, Wilson EO (1967) The theory of island biogeography. Princeton University Press, Princeton, NJ
MacDonald IAW, Loope LL, Usher MB, Hamann O (1989) Wildlife conservation and the invasion of nature reserves by introduced species: a global perspective. In: Drake JA, Mooney HA, de Castri F, Graves RH, Kruger FJ, Rejmanek M, Williamson M (eds) Biological invasions: a global perspective. Wiley, New York, pp 215–256
McInnes PF, Naiman RJ, Pastor J, Cohen Y (1992) Effects of moose browsing on vegetation and litter of the boreal forest, Isle Royale, Michigan, USA. Ecology 73: 2059–2075
McIntosh RP (1981) Succession and ecological theory. In: West DC, Shuggart HH, Botkin DB (eds) Forest succession: concepts and applications. Springer, Berlin Heidelberg New York, pp 227–304
Mead AR (1992) Two giant African land snail species spread to Martinique, French West Indies. Veliger 35: 74–77
Molvar EM, Bowyer RT, Van Ballenberghe V (1993) Moose herbivory, browse quality, and nutrient cycling in an Alaskan treeline community. Oecologia 94: 472–479
Mooney HA, Drake JA (1986) Ecology of biological invasions of North America and Hawaii. Springer, Berlin Heidelberg, New York
Naiman RJ (1988) Animal influences on ecosystem dynamics. Bioscience 38: 750–752
Naiman RJ, Johnston CA, Kelley JC (1988) Alterations of North American streams by beaver. Bioscience 38: 753–762
O'Dowd DJ, Lake PS (1989) Red crabs in rain forest, Christmas Island: removal and relocation of leaf-fall. J Trop Ecol 5: 337–348
O'Dowd DJ, Lake PS (1990) Red crabs in rain forest, Christmas Island: differential herbivory of seedlings. Oikos 58: 289–292
O'Dowd DJ, Lake PS (1991) Red crabs in rain forests, Christmas Island: removal and fate of fruits and seeds. J Trop Ecol 7: 113–122

Paine RT (1966) Food web complexity and species diversity. Am Nat 100: 65–75
Paine RT (1969) A note on trophic complexity and community stability. Am Nat 91–93
Pastor J, Naiman RJ, Dewey B, McInnes P (1988) Moose, microbes, and the boreal forest. Bioscience 38: 770–777
Pastor J, Dewey B, Naiman RJ, McInnes PF, Cohen Y (1993) Moose, browsing and soil fertility in the boreal forests of Isle Royale National Park. Ecology 74: 467–480
Robertson AI (1991) Plant-animal interactions and the structure and function of mangrove forest ecosystem. Aust J Ecol 16: 433–443
Rodda GH, Fritts TH (1992) The impact of the introduction of the colubrid snake Boiga irregularis on Guam's lizards. J Herpetol 26: 166–174
Savige JA (1987) Extinction of an island forest avifauna by an introduced snake. Ecology 68: 660–668
Schulze E-D, Mooney HA (1993) Biodiversity and ecosystem function. Springer, Berlin Heidelberg New York
Siegfried WR, Condy PR, Laws RM (1985) Antarctic nutrient cycles and food webs. Springer, Berlin Heidelberg New York
Smith FDM, May RM, Pellwe R, Johnson TH, Walker KR (1993) How much do we know about the current extinction rate? Trends Ecol Evol 8: 375–378
Smith VR (1978) Animal-plant-soil nutrient relationships or Marion Island (Subantarctic). Oecologia 32: 239–253
Smith VR, Steenkamp M (1990) Climatic change and its ecological implications at a sub-antarctic island. Oecologia 85: 14–24
Thompson JN (1988) Variation in interspecific interactions. Annu Rev Ecol Syst 19: 65–87
Vitousek PM (1986) Biological invasions and ecosystem properties: can species make a difference? In: Mooney HA, Drake JA (eds) Ecology of biological invasions of North America and Hawaii. Springer, Berlin Heidelberg New York, pp 163–178
Vitousek PM (1990) Biological invasions and ecosystem processes: toward an integration of population and ecosystem studies. Oikos 57: 7–13
Vitousek PM, Hooper DU (1993) Biological diversity and terrestrial ecosystem biogeochemistry. In: Schulze E-D, Mooney HA (eds) Biodiversity and ecosystem function. Springer, Berlin, Heidelberg New York, pp 3–14
Vitousek PM, Walker LR (1989) Biological invasion by *Myrica faya* in Hawaii: plant demography, nitrogen fixation, and ecosystem effects. Ecol Monogr 59: 247–265
Vitousek PM, Walker LR, Whiteaker LD, Mueller-Dombois D, Matson PA (1987) Biological invasion by *Myrica faya* alters ecosystem development in Hawaii. Science 238: 802–804
Vivrette NJ, Muller CH (1977) Mechanism of invasion and dominance of coastal grassland by *Mesembryanthemum crystallinum*. Ecol Monogr 47: 301–318
Walker BH (1992) Biodiversity and ecological redundancy. Conserv Biol 6: 18–23
Woodward FI (1993) How many species are required for a functional ecosystem? In: Schulze E-D, Mooney HA (eds) Biodiversity and ecosystem function. Springer, Berlin Heidelberg New York, pp 271–291

12 Biological Diversity and the Maintenance of Mutualisms

D.R. GIVEN

12.1 Ecological Interactions

In most ecosystems it is species which are the initial focus; but the fundamental dynamics of the system are expressed in the interactions between populations of species making up the community rather than the species themselves.

The principal kinds of ecological interactions between species are competition, predation and mutualism. All are population phenomena in which the effect of one species on the population growth rate and/or population size can be regarded as positive or negative. In the simplest model, a competitive interaction is indicated by negative effects on component species. Where there is a benefit for one species but a corresponding negative interaction for others, predation is indicated. If there are positive outcomes all round, the interaction type is mutualistic.

Boucher (1985a) provides a very simple definition of mutualism as being where "different kinds of organisms help each other out. This, in brief, is the idea of mutualism". An evolutionary-based definition is given by Janzen (1985) as "an interaction between individual organisms in which the realised or potential genetic fitness of each participant is raised by the actions of the others". An important point stressed by Janzen is that there is a delicate balance and the mutualistic relationship of today can change to become the parasitism of tomorrow.

Mutualisms can be conveniently divided into facultative and obligatory, and into diffuse and one-on-one mutualisms. Janzen (1985) suggests that the most usual case is diffuse and apparently facultative; the removal of one partner in the relationship does not necessarily result in the extinction of the other. Completely obligate examples of mutualisms such as that suggested between the Australian shrubs *Verticordia nitens* and *V. aurea*, and oil-ingesting bee pollinators (Houston et al. 1993) are probably quite unusual. Janzen recognized five types, each of which has relevance to islands. These are harvest mutualisms, pollination mutualisms, protective mutualisms, seed dispersal mutualisms and human agriculture or husbandry.

Mutualism is generally less extensively studied than either competition or predation. Pollination by insects, invertebrate relationships and dispersal of

Department of Plant Science, Lincoln University, P.O. Box 84, Lincoln University, Canterbury, New Zealand

Ecological Studies, Vol. 115
Vitousek et al. (eds) Islands

fruits and seeds by a wide range of animals are the most commonly cited types of direct mutualism; but other types may, as Boucher suggests, be more common than currently available data indicates. Janzen (1985) suggests that mutualisms, "are the most omnipresent of any organism-to-organism interaction" and that, "all terrestrial plants, vertebrates, and arthropods are involved in one diffuse mutualism and many are involved in several". It has been suggested that mutualisms may be difficult to observe rather than rare (Wolin 1985).

12.2 Representative Examples of Mutualisms

12.2.1 Ant-Plant Relationships

One of the best-documented types of mutualism concerns obligate relationships between ants and plants, or myrmecophily. Every major tropical landmass, and to a lesser extent islands and temperate regions, has at least a few obligate ant-plants in which the plant provides living space and food for ants. In exchange for this the ant protects the plant from potential predators and choking vines. Studies of such a relationship are provided by Janzen (1972) on protection of the small tree *Barteria* in Nigeria by *Pachysima* ants, McKey (1984) on *Leonardoxa africana* and the mutualist ant *Petamomyrmex phylax* in Cameroon, Longino (1989) on the relationships between *Cecropia* and *Azteca* ants in Costa Rica, Kleinfeldt (1978) on ant gardens, and Wilson (1987) on arboreal ant faunas in Peru.

Longino (1989) found a difference in competitive ability resulting in non-obligate ants never becoming dominant in *Cecropia* trees. Explanation may lie in fundamental differences in the nesting behaviour of obligate *Azteca* ants and non-obligate ants. Whereas for the former the entire tree forms the potential nest boundary and nests continually expand, for non-obligate species colonies are either small, or are large but spread in many small nests. Such competitive ability can be a critically important factor in determining resilience of island biota and the establishment success of immigrants.

Collapse of an ant-plant mutualism through competition with an invading species has been documented for *Mimetes cucullatus* (Proteaceae), a member of the unique southern African fynbos flora (Bond and Slingsby 1984). Indigenous ants are important seed dispersers, taking seeds for food and storing them in their nests where a proportion germinate. Invasion by the Argentine ant, *Iridomyrmex humilis*, has resulted in a dramatic reduction in indigenous ant numbers. Whereas seedling emergence of *Mimetes* was 35% in sites not infested by *Iridomyrmex*, this dropped to 0.7% in infested sites. Bond and Slingsby predict that continued invasion of fynbos by *Iridomyrmex* may eventually lead to marked loss of biodiversity by extinction of many rare, endemic Cape Proteaceae by slow and subtle attrition of seed reserves.

Another important consideration for island biology is the vulnerability of islands to disturbance which can have profound effects upon population density, dispersion and patterns of ant reproduction. This has been demonstrated for *Sanguinaria canadensis* (Pudlo et al. 1980) for which ants are a major seed dispersal vector. Whereas seed dispersal was high at undisturbed sites, on the most disturbed sites seed dispersal virtually ceased. Although this resulted in high densities of *Sanguinaria* where disturbance was greatest, it also promotes increasing inbreeding, as seeds are rarely relocated beyond the parent clone.

Island-based evolutionary studies of ant-plant relationships include those of Janzen (1973) and Rickson (1977) for selected Carribean islands. Janzen notes that *Cecropia peltata* is not occupied by *Azteca* ants on most islands of the Caribbean and that this is associated with loss of trichilia which produce glycogen-rich food for ants. Rickson's study sampled selected populations of *Cecropia peltata* from mainland South American populations and islands between Trinidad and Guadaloupe. The results showed that production of Mullerian bodies from which ants feed decreases rapidly in the West Indies north of Trinidad and Tobago. On Puerta Rico *Cecropia* has lost all structures normally associated with the production of ant food. Initial losses are in the number of Mullerian bodies, and subsequently production ceases from the remaining bodies, sequentially followed by reduction in length of associated trichomes.

Both Janzen and Rickson argues that forms of *Cecropia* without ants are descended from those with ants. Rickson suggests that this is a result of relaxed selection pressures from insects and vines, and that the production of food for ants is an expensive business which if it can be avoided allows the plant to channel more of its energies into increased growth and reproduction. Janzen argues for reduced herbivore pressure in insular habitats, as well as reduced interspecific competition from trees in those island habitats that are poor in vines. He notes that there is similar loss of mutualistic ant-plant interactions at higher elevations on mainland sites, and that complex mutualisms can disintegrate without the loss of partners if the proper habitats are available.

Present ant-plant associations may sometimes reflect past biological constraints; Janzen (1972) argues from indirect evidence that the relationship between *Barteria* and *Pachysima* may have developed in the past when large browsing mammals were a threat to *Barteria*. He suggests that this is in contrast to the neotropics, which lacks evidence that browsing mammals were of major importance in the coevolution of ant-plant mutualisms. Relict mutualistic associations should be considered especially for those island systems such as Madagascar, New Zealand and New Caledonia, which are fragments of former continents.

12.2.2 Ants and Other Invertebrates

Ants interact not only with plants but also with a variety of other invertebrates including homopterans (e.g., Buckley 1990) and lycanid butterfly larvae (e.g., Pierce and Mead 1981). The presence of ants often increases the survival of homopterans which can have deleterious effects on the plant host (e.g., Messina 1981). However, the ants can exclude other herbivores and if this outweighs damage caused by homopterans then an indirect mutualism is established in which all parties benefit (Compton and Robertson 1988), the outcome of a particular interaction being affected by a variety of factors. These are not only biotic, for instance, abundance of participating species, but also abiotic factors such as fire, soil and climate.

Differences in ant aggressiveness have not been generally noted in ant-plant-invertebrate mutualisms but this is indicated in a study of interactions between 11 ant species near Madang, Papua New Guinea (Buckley and Gullan 1991). They found that greater protection of homopterans from parasitism was provided where ants were more aggressive. This has important implications where islands are invaded by more aggressive ant species, as has occurred in many parts of the tropics, especially if the exotic species have potential to displace those that are indigenous.

There is at least one example from the Solomon Islands where native and introduced ants are differentially effective in attacking Hemiptera on coconut palms (*Cocos nucifera*) rather than tending them (Greenslade 1971). In this example, *Oecophylla smaragdina* (indigenous) and *Anoplolepis longipes* (introduced) protect coconuts against the coreid hemipteran, *Amblypelta cocophaga*, whereas *Iridomyrmex cordatus* (indigenous) and *Pheidole megacephala* (introduced) do not. Initial violent oscillations in the relative diversity of the ants has gradually moderated. Greenslade suggests that further major frequency changes, especially increases in the two exotic species, are possible, if not probable, under plantation conditions where low biotic diversity contrasts with the high diversity of indigenous forest.

Research on ant-plant-lepidopteran systems, which has yet to be matched by parallel studies of ant-plant-homopteran systems, suggests that proximity to ant nests may be important in differential selection of host plants for occupation. This in turn relates to patterns of nesting behaviour and ant nest size, as well as patterns of dispersal when ants invade new sites.

12.2.3 Figs and Insect Pollinators

The 900-odd species of *Ficus* constitute one of the most distinctive genera of tropical plants, of extraordinary diversity and wide distribution, both continental and oceanic. One of the more remarkable features of the group is its total reliance on complex obligatory mutualisms with pollinating agaonid wasps. The figs and wasps are interdependent and virtually every species of fig has its own species of pollinator wasp (Wiebes 1979). The ostiole (the entry point into the fig flower) is generally considered to act as a filter which prevents non-adapted wasps from entering the "wrong" figs. Janzen (1979) points out that figs demonstrate extreme intra-population inter-tree asynchronous flowering and fruiting in many habitats, yet strong intra-tree synchronous flowering and fruiting. This means that even rare figs are located and pollinated at very low densities with extreme efficiency by fig wasps.

The colonization of islands by figs and their wasps contains a relevant paradox (Janzen 1979). A single seed from a mainland cannot start a fig population because the single tree it produced cannot sustain a pollinating wasp population. Likewise, a pollinating wasp population cannot survive until there is a population of fig trees. Yet most tropical islands possess one or more species of *Ficus* (with the notable exception of the Hawaiian archipelago). Janzen suggests that the most likely solution is the extension of a mainland fig-seed shadow to an island by fruit pigeons or bats followed by wasp colonization of the resultant island population of the mainland fig genome.

An obvious question is whether there is a breakdown in flowering synchrony or a shift in wasp specificity. Observation of figs on small San Andres Island (Colombia) suggests that such a breakdown in synchrony can occur (Ramirez 1970). The few trees of two *Ficus* species on San Andres usually had syconia in all phases of development. Similarly, there is some evidence that under very harsh mainland environments similar to some islands, agaonid wasps can evolve to pollinate two or more species (Janzen 1979).

Breakdowns in pollinator specificity in *Ficus* have been recorded (e.g., Ware and Compton 1992). They describe successful pollination of a single plant of *Ficus lutea* growing 500 km outside its normal distribution. Small numbers of agaonid wasps which normally pollinate two other figs entered and successfully pollinated this species. The alien species were not initially attracted to *F. lutea* but, having landed, were stimulated to locate the ostiole and enter the fig cavity. Females of a third fig wasp were also present but failed to initiate ostiole searching. This suggests potential for pollinator switching involving island immigrants provided alternative agaonid wasps are present.

Buckley (1987) notes the paucity of studies on mutualisms involving more than a single pair of participants. One example is the need to consider spider predation and parasitoid wasps as part of ant-plant homopteran systems (Buckley 1990). *Ficus* provides another interesting example of a more complex situation (Compton and Robertson 1988). They show that a mutualism between ants and homopterans can benefit the mutualism between *Ficus* species and their obligate Agaonidae wasp pollinators. Eggs of the pollinating wasps are prone to parasitism by *Apocrypta guineensis*, a fig wasp parasitic on larvae of other species. By excluding ants they demonstrated that ant presence can have a dramatic effect on the degree of parasitism of fig pollinating wasps and on the extent of seed destruction by seed predating wasps. In this example at least six invertebrates and *Ficus* interact in a complex series of relationships.

12.2.4 Bat Pollination Systems

Plant pollination systems involving vertebrates (especially ornithophily involving birds and chiropterophily involving bats) have been well studied as examples of mutualisms, the plant providing food and the vertebrate providing a means of achieving outcrossing. One of the problems facing island plants is that there is a general tendency for diversity of vertebrates (with the exception of birds) to decrease inversely with oceanicity. This can create problems for plants which have an obligate relationship with vertebrates. The very low diversity of animal pollinators on oceanic islands can act as a potential biotic filter to plant immigrants (Elmquist et al. 1992).

Of particular interest for studies of mutualisms on oceanic islands are pollination systems involving bats. In the South Pacific bat, diversity sharply decreases with increasing distance from continents. Study of the pollination of kapok (*Ceiba pentandra*) by bats shows that whereas in continental areas it attracts a wide assemblage of pollinators, in Samoa it is pollinated by only one single bat, *Pteropus tonganus* (Elmquist et al. 1992). That pollination even occurs is a reflection of the generalist traits of both plant and bat. The reliance of *C. pentandra* on *P. tonganus* carries a cost in up to 50% of flowers and initiated fruit being damaged, although in other respects the bat is a reasonably efficient pollinator.

Parallel examples of particular relevance to islands concern the genus *Freycinetia* (Pandanaceae). This genus includes dioecious vines with inflorescences which are large, showy congested spikes subtended by fleshy bracts. The genus is widely distributed through the tropics, especially the islands of the Pacific region. It provides good examples of pollination mutualisms involving both birds and bats. The attractant is not nectar, the principal rewards being the fleshy bracts and the waxy pollen matrix from staminate spikes.

There is evidence that pollinator flexibility and opportunism assist *Freycinetia* to colonize new islands where any particular component of the vertebrate fauna may be absent. Study of *F. reineckei* on Samoa (Cox 1984) shows that this particular species is both ornithophilous and chiropterophilous and that the mutualism is both opportunistic and non-coevolutionary. There is no specific adaptation for visitation by the vertebrate pollinators which feed on *F. reineckei* during the flowering period (March through May) but they feed on a variety of other fruits and possibly flowers during the rest of the year.

Freycinetia arborea of Hawaii provides a second example of a flexibility in a pollination mutualism in an island context. In this instance the relationship has changed during historic time (Cox 1983). Although for a long time this species was believed to be pollinated by rats, Cox provides firm evidence for pollination by endemic Hawaiian birds of the Drepanididae and Corvidae). This comes from early ornithological observations as well as analysis of feathers of birds of these groups for *Freycinetia* pollen. Within the last century, near-extinction and extinction of these birds has resulted in a shift to other pollinators, principally *Zosterops japonica*, the silver eye, introduced to Hawaii from Japan in 1929.

There are two particularly interesting aspects of the example of *Freycinetia arborea*. One is that there may have been a period of 15 to 20 years during which *Freycinetia arborea* was only sparsely pollinated. However, this species is capable of vigorous clonal growth and Cox suggests that it is unlikely that this temporary hiatus in pollination severely reduced the range of *Freycinetia*. He does suggest that it may, however, have favoured any bisexual individuals in the population.

This is one of the few examples of an indigenous species experiencing a complete change from an endemic to an introduced pollinator. Cox points out that the original pollination system was unlikely to be coevolved. The accessibility of the pollination system to a wide range of potential pollinators indicates its unspecialized nature. This contrasts with tightly evolved ornithophilous systems frequently found in the neotropics.

A remarkable instance of the potential for biological disturbance and imbalance to raise the possibility of extinction on an island concerns the recent discovery of bat-plant interactions on the North Island of New Zealand (Ecroyd 1993). This involves the endemic root parasite *Dactylanthus taylori*, a monotypic genus which is the only indigenous member of the Balanophoraceae, and the short-tailed bat (*Mystacina tuberculata*), which is not only endemic but, with one other species, belongs to a distinct family. An obligate mutualistic relationship appears to be involved with the bats travelling long distances to feed on the large quantities of nectar produced by flower clusters. This relationship has probably developed over many millions of years since separation of the New Zealand archipelago from Gondwana. Both partners are now highly threatened species and instead of bats, the flowers of *Dactylanthus* are now attracting introduced feral mammals such as Australian possums and Polynesian rats, which destroy the flowers and prevent seed production.

12.3 Chatham Islands: Ecosystem Disruption and Mutualisms

12.3.1 *Sophora microphylla* Pollination

Sophora microphylla, or kowhai, a small tree, is widespread throughout New Zealand. It is genetically variable and although only three taxa have been formally recognised at infraspecific level, plants from particular geographic regions are often readily distinguished from those of other regions.

The Chatham Islands lie east of the New Zealand mainland. They have been greatly modified since European settlement in the 19th century with a high degree of ecosystem fragmentation and have been subjected to an extraordinary number of bird extinctions. *Sophora* here is restricted to sites on limestone along the western margin of Te Whanga Lagoon. Some populations are in poor general health and there is a general scarcity of young plants and seedlings. Investigation of seed/ovule (S/O) ratios has indicated that for the Chatham Island these ratios are comparatively low in contrast to the ratio for a mainland plant of Chatham provenance (Table 12.1).

In considering the reasons for low seed-set on the Chatham Islands, and nothing especially that the mainland tree of Chatham Island origin has a relatively high S/O ratio, there are two significant points. The first is that kowhai produces an abundance of nectar and would have been frequently visited in the past as a food source by Chatham Island bellbird (*Anthornis melanura melanocephala*) and Chatham Island tui (*Prosthemadera novaeseelandiae*

Table 12.1. Variable seed-ovule Ratios in *Sophora microphylla* from Chatham Islands

Site[a]	No. of pods	Potential seed/pod	Viable seed/pod	S/O ratio
Chatham Is. sites				
1	45	3.4	0.8	26%
2	7	5.3	1.0	19%
3	37	8.5	2.7	32%
4	28	5.8	3.5	60%
Mainland ex situ site				
5	30	8.1	6.5	80%

[a] Site details: 1 = Blind Jims Creek, north end of population; 2 = Blind Jims Creek, south end of population; 3 = north of Te Matarau Point; 4 = south of Te Matarau Point; 5 = cultivated tree ex Chatham Is, DSIR Lincoln.

chathamensis). The Chatham Island bellbird is probably extinct and the Chatham Island tui is very scarce and is probably endangered. No effective substitute species have immigrated to these islands.

The second point is that variations in fruit set in *Sophora microphylla* suggest that there are differences in the extent of self-pollination which can occur (Rattenbury 1979). Rattenbury hypothesised that this is regulated by the time of style elongation before anthesis, and could be accounted for by a single gene difference. A breeding system which ensures a continuing supply of seed independent of availability of pollen vectors, but also permitting some out-crossing, should be favoured. Under normal circumstances this would greatly favour high seed-producing homozygous plants rather than seedless homozygous plants or heterozygotes.

Sophora on the Chatham Islands may be an example of a species which had high seed production even from heterozygotes when the tui and bellbird were in reasonably high numbers on the Chatham Islands. At this time, there would be little selection against self-incompatibility. Now, however with impeded ability to cross-pollinate, it is likely that the resulting increase in self-pollination lacks efficiency because the proportion of self-incompatible plants is unusually high in some populations on the Chatham Islands.

12.3.2 *Rhopalostylis* Seed Dispersal

Observations at Nikau Bush Reserve and Lake Huro show that once mature, the endemic *Rhopalostylis*, nikau palm, flowers and fruits heavily. The seed is heavy and drops directly beneath the crown of fronds. A substantial seed bank, up to 10 cm in depth, is built up under mature adults. Little germination of the fallen seed in the seed bank appears to take place except in constantly damp ground underlying fallen palm fronds, or under dense and well shaded undergrowth.

Most groups of young plants and seedlings do not occur immediately under adult palms but up to 200 m away. Along three transects, regeneration of *Rhopalostylis* is highly contagious, with highest concentrations of seedlings and young plants generally directly below plants of *Pseudopanax chathamicus*, Chatham Island hoho. This tree is favoured by Chatham Island pigeon (*Hemiphaga novaeseelandiae* subsp. *chathamica*), both for food and as a roost. A likely explanation is that pigeons feed on nikau, and void the seeds while roosting, as noted by McEwan (1978) for mainland pigeons. On the Chatham Islands nikau fruits constitute an important seasonal food. (K. Hughey, pers. comm. 1991).

The Chatham Island Pigeon is now endangered. The last pair known to regularly frequent Nikau Bush disappeared from the area about 1986–87 (G. Murman, pers. comm. 1991). This correlates with a marked regeneration gap between tall mature adults, and seedlings of *Rhopalostylis*. In Nikau Bush Reserve, the prominent class of seedling nikau plants probably dates from fencing of the reserve in the early 1980s and exclusion of stock.

Opportunity for the role of pigeons to be taken over by introduced Australian brush-tailed possums is limited. Possums have difficulty climbing mature palm trunks, although they eat both flowers and fruit, given the opportunity (J. Coleman, pers. comm. 1991). Individual possums range widely, voiding seeds in their faecal pellets as they go, so that they would give rise to a diffuse pattern of seedlings.

12.4 Seabirds and Soil–Commensalisms or Indirect Mutualisms

In the New Zealand region, perhaps 500 000 fairly prions (*Pachyptila turtur*) return each spring to Stephens Island with an area of only 150 ha. Their faeces enrich the soils, giving rise to an abundant invertebrate community that supports populations of seven species of lizard, one frog, and 50 000 tuatara. This points to one of the most remarkable features of primeval New Zealand. Prior to the coming of humans about 1000 years ago seabirds at high densities were keystone species that supported and maintained high biological diversity of ecosystems (Daugherty et al. 1990).

Naturalists have noted for over a century that throughout its range the tuatara (*Sphenodon punctatus*) has an intimate relationship with burrowing seabirds (e.g., Walls 1978). However, the ecological relationships are still poorly understood, for instance whether they are purely commensalistic or mutualistic in part. Tuatara have been postulated to be dependent on prions and other seabirds, and tuataras not only use the burrows of the birds but also prey on chicks and injured adults; but energy flows have not been studied in detail. What is appreciated is that there exists a delicate balance which is readily upset. Studies by Crook (1973) have, for instance, established that tuataras cannot persist on those islands which have been invaded by rats (including the Polynesian rat, *Rattus exulans*).

The New Zealand subantarctic islands also preserve strong sea-bird influence. On Campbell Island to the south of the New Zealand mainland, a narrow but extensive zone of biotic soils has been mapped along much of the island's littoral zone (Meurk and Given 1990). This is enriched by the faeces of marine mammals and seabirds and gives rise to a distinctive vegetation system featuring genera such as *Leptinella* and *Crassula*. On the Chatham Islands further north, species such as *Leptinella featherstonii* and *Myosotidium hortensia* are restricted to such nutrient-rich littoral soils. On the New Zealand mainland, where seabird and marine mammal numbers are now lower, it is significant that coastal plants such as *Lepidium* spp. are threatened rare to extinct, and recently Garnock-Jones and Norton (1995) have suggested that decline in sea birds may be one of the most significant causes of depletion.

Although none of the examples is yet a clearly proven case of mutualism, it is highly unlikely that benefits to various partners will be one-way. Investigation of interactions within these complex systems is in its infancy, but does have a measure of urgency. The impact of humanity in these island systems has resulted in a dramatic decline in seabird numbers, and has the potential to further alter greatly, and perhaps irreversibly, the relationships between indigenous species of animals and plants.

12.5 The Survival of Mutualisms on Islands

Obligatory mutualisms are likely to weaken as one progresses from continental regions to off-shore islands and then onto more distant islands (e.g., Rickson 1977). On younger islands, mutualisms will probably be more generalist, in a state of development and adjustment, and opportunistic. Major exceptions should be on older islands with relict continental biotas such as Madagascar, New Caledonia and the New Zealand archipelago. In common with other aspects of island biology, these are likely to be highly vulnerable to human-induced changes in ecosystems and habitats.

An important question is whether replacement or substitute species (which may be exotic invaders) can maintain a mutualistic relationship once one of the original partners disappears. An allied question is whether island populations expand food sources and occupied habitat or have realised broader niches that on continents (Feinsinger and Swarm 1982). As these authors point out, the matter is complicated in that it appears to be only during periods of resource (e.g., food) shortage that the realized niche reflects the capacity of the population's phenotype.

Replacement has been demonstrated in some island contexts (e.g., Cox 1983; Ware and Compton 1992), but examples are few at present. Study of ant-plant relationships in the orchid *Spathoglottis plicata* (Jaffe et al. 1989) demonstrates the potential. Originally from Borneo, the orchid is new widely distributed in the neotropics. On Guadaloupe it is subject to ant foraging by several species which are attracted by sweet floral secretions. These are different species from those ants which are associated with the plant in Borneo, and in their relationships with the *Spathoglotis* they form a gradient of "beneficial species".

Analysis of an interspecific interaction or class of interactions should ideally consider: (1) its effects on the population dynamics, behaviour and energy/nutrient budgets of each participant; (2) the costs and benefits of the interaction to each participant, and whether these are similar for each participant or dissimilar; (3) how taxon-specific, and how obligate or facultative it is for each participant; and (4) what morphological, behavioral and physiological specialisations are involved.

Buckley (1987) points out that these aspects have been examined with various degrees of precision for a range of twofold interactions and a more limited range of higher-order interactions. Such higher-order interactions involving three or more participants form an important field for investigation because they represent closer approximations to real ecosystems than do twofold systems. Higher-order studies also have a particular relevence for islands where there are likely to be few surrogate species which can replace lost partners in biotic relationships.

The need to place greater emphasis on the study of higher order interactions is demonstrated by the question of the place of parasitism in biological systems. Ecosystems with high species *diversity* are usually assumed to have a high level of stability; but increasing *complexity* (the degree of interconnection among system components) leads to inherent instability in systems (Watt et al. 1977; Gilbert 1980). Freedland and Boulton (1992) argue that such systems are probably stabilised by coevolved food web interactions in which parasitism may play a vital role. They suggest that omnivores and mutualists destabilize food webs without parasites. If this is the case, there may need to be significant changes in management of ecosystems and development of new conservation practices outside current systems of island national parks and reserves.

Considering reproduction as vital to the survival of species and hence maintenance of biodiversity, Cox et al. (1991) conclude that the dependency of highly endemic island floras on few potential pollinators in depauperate island faunas suggests that pollinators and seed dispersers may be crucial in the preservation of biodiversity in isolated oceanic islands. They point out that continued decline and ultimate extinction of flying fox species on Pacific Islands may lead to a cascade of linked extinctions. Bond and Slingsby (1984) argue similarly for the fate of some proteaceous members of the fynbos flora of southern Africa in the face of invasion by argentine ants.

If many forested tropical islands follow the generalised pattern for lowland tropical rainforests generally (Bawa 1990), they should be experiencing intense selection for long-distance pollen flow. Bawa has estimated that approximately 98–99% of all flowering plant species in tropical lowland rainforest are pollinated by animals. However, it is possible that generally low numbers of indigenous mammals on many islands, coupled with lesser predictability of weather events such as irregular cyclonic storms, results in less intense selection and more conservative pollination scenarios.

12.6 Management Under a Monoculture Scenario

Janzen (1985) argues that a world without mutualisms would be one of narrow ecological ranges, more resources required for reproduction and to counter the vagaries of competition, greater edaphic specialisation, and restriction to "good" soils (or minimal limitation to essential resources). He predicts that a non-mutualistic world would see a greater number of monospecific stands with lower within-habitat diversity and an array of habitat specialists and incompetent generalists. To quote Janzen (1985): "I suppose one could invent a plant phenotype that walked about planting its seeds in little piles of fertilizer in forest tree falls, but it certainly involves many fewer steps and resources to surround the tree with a bit of nutrient tissue and thereby splice a forest bison to your genome".

The scenario pictured by Janzen is reminiscent of many of the human-generated ecosystems found around the world, and indeed, Boucher (1985b) suggests that: "One way to aid domesticated organisms has been to aid their mutualists.... However, an alternative path has also been followed, and indeed seems to be the predominant trend in agriculture in recent years. This is simply to eliminate the mutualists and replace them with direct human intervention".

Ex situ conservation may not provide a simple answer to potential biotic collapse through breakdown in the inter-relationships between groups of species. To cite one example, study of cycads shows that they have a large number of associated animals, including obligate pollinators. Many cycads are highly threatened in the wild state; but maintaining and propagating living collections of cycads with eventual reintroduction will not work unless the associated fauna is also preserved (Vovides 1991).

Overall, perhaps the greatest problem is paucity of definitive information on and modelling of island mutualisms. Instances such as the very recent discovery in New Zealand of a mutualistic relationship between endangered bats and an endangered root parasite, *Dactylanthus taylorii* (Ecroyd 1993) demonstrate the simple observations that remain to be made.

Whether the preservation of island mutualisms is important depends on what we collectively consider to be the key components of the biodiversity ark of the 21st century. Are they related only to preservation of representative segments of taxonomic species? If the objective is to also preserve genetic diversity, evolutionary potential, and systems and processes then investigation and facilitating of mutualisms must be incorporated into any recovery or conservation programme.

Acknowledgments. I am grateful to Professor Peter Vitousek for the invitation to prepare this chapter and to Dr David Norton and Professor Henning Adsersen for reviewing it. Thanks are due to the many colleagues with whom I have discussed issues in island biology and whose ideas have shaped my own.

References

Bawa KS (1990) Plant-pollinator interactions in tropical rain forests. Annu Rev Ecol Syst 21: 399–422
Bond W, Slingsby P (1984) Collapse of an ant-plant mutualism: the Argentine ant (*Iridomyrmex humilis*) and myrmecochorous proteaceae. Ecology 65: 1031–1037
Boucher DH (1985a) The biology of mutualism. Croom Helm, London
Boucher DH (1985b) Mutualisms in agriculture. In: Boucher DH (ed) The biology of mutualism. Croom Helm, London, pp 375–386
Buckley R (1987) Ant-plant-homopteran interactions. Adv Ecol Res 16: 53–79
Buckley R (1990) Ants protect tropical Homoptera against nocturnal spider predation. Biotropica 22: 207–209
Buckley R, Gullan P (1991) More aggressive ant species (Hymenoptera: Formicidae) provide better protection for soft scales and mealybugs (Homoptera: Coccidae, Pseudococcidae). Biotropica 23: 282–286
Compton SG, Robertson HG (1988) Complex interactions between mutualisms: ants tending homopterans protect fig seeds and pollinators. Ecology 69: 1302–1305
Cox PA (1983) Extinction of the Hawaiian avifauna resulted in a change of pollinators for the 'ie'ie, *Freycinetia arborea*. Oikos 41: 195–199
Cox PA (1984) Chiropterophily and ornithophily in *Freycinetia* (Pandanaceae) in Samoa. Plant Syst Evol 144: 277–290
Cox PA, Elmquist T, Pierson ED, Rainey WE (1991) Flying foxes as strong interactors in South Pacific island ecosystems: a conservation hypothesis. Conserv Biol 5: 448–454
Crook IG (1973) The Tuatara, *Sphenodon punctatus* Gray, on islands with and without populations of the Polynesian rat, *Rattus exulans* (Peale). Proc NZ Ecol Soc 20: 115–120
Daugherty CH, Towns DR, Atkinson IAE, Gibbs GW (1990) The significance of the biological resources of New Zealand islands for ecological restoration. In: Towns DR, Daugherty CH, Atkinson IAE (eds) Ecological Restoration of New Zealand Islands. New Zealand Department of Conservation Sciences Publ 2, Wellington pp 9–22
Ecroyd C (1993) In search of the wood rose. Forest and Bird Febr 1993: 24–28
Elmquist T, Cox PA, Rainey WE, Pierson ED (1992) Pollination of *Ceiba pentandra* by flying foxes in Samoa. Biotropica 24: 15–23
Feinsinger P, Swarm LA (1982) "Ecological release", seasonal variation in food supply, and the hummingbird *Amazilia tobaci* on Trinidad. Ecology 63: 1574–1587
Freedland WJ, Boulton WJ (1992) Coevolution of food webs: parasites, predators and plant secondary compounds. Biotropica 24: 309–327
Garnock-Jones PJ, Norton DA (1995) *Lepidium naufragorum* (Brassicaceae), a new species from Westland, and notes on other New Zealand coastal species of *Lepidium*. NZ J Bot 33: (in press)
Gilbert LE (1980) Food web organization and conservation of neotropical diversity. In: Soule ME, Wilcox BA (eds) Conservation biology. Sinauer, Sunderland, MA, pp 11–34
Greenslade PJM (1971) Interspecific competition and frequency changes among ants in Solomon Islands coconut plantations. J Appl Ecol 8: 323–352
Houston TF, Lamont BB, Radford S, Errington SG (1993) Apparent mutualism between *Verticordia nitens* and *V. aurea* (Myrtaceae) and their oil-ingesting bee pollinators (Hymenoptera: Colletidae). Aust J Bot 41: 369–380
Jaffe K, Pavis C, Vansuyt G, Kermarrec A (1989) Ants visit extrafloral nectaries of the orchid *Spathoglotis plicata* Blume. Biotropica 21: 278–279
Janzen DH (1972) Protection of *Barteria* (Passifloraceae) by *Pachysima* ants (Pseudomyremecinae) in a Nigerian rain forest. Ecology 53: 885–892
Janzen DH (1973) Dissolution of mutualism between *Cecropia* and its *Azteca* ants. Biotropica 5: 15–28
Janzen DH (1979) How to be a fig. Annu Rev Ecol Syst 10: 13–51
Janzen DH (1985) The natural history of mutualisms. In: Boucher DH (ed) The biology of mutualisms. Croom Helm, London, pp 40–99

Kleinfeldt SE (1978) Ant-gardens: the interaction of *Codonanthe crassifolia* (Gesneriaceae) and *Crematogaster longispina* (Formicidae). Ecology 59: 449–456
Longino JT (1989) Geographic variation and community structure in an ant-plant mutualism: *Azteca* and *Cecropia* in Costa Rica. Biotropica 21: 126–132
McEwan WM (1978) The food of the New Zealand pigeon (*Hemiphaga novaeselandiae novaeseelandiae*). NZ J Ecol 1: 99–108
McKey D (1984) Interaction of the ant-plant *Leonardoxa africana* (Caesalpiniaceae) with its obligate inhabitants in a rain forest in Cameroon. Biotropica 16: 81–99
Messina FJ (1981) Plant protection as a consequence of an ant-membracid mutualism: interactions on goldenrod (*Solidago* sp.). Ecology 62: 1433–1440
Meurk CD, Given DR (1990) Vegetation map of Campbell Island. Scale 1: 25,000. Map (1 sheet). DSIR Land Resources, Christchurch
Pierce NE, Mead PS (1981) Parasitoids as selective agents in the symbiosis between lycaenid butterfly larvae and ants. Science 211: 1185–1187
Pudlo RJ, Beattle AJ, Culver DC (1980) Population consequences of changes in an ant-seed mutualism in *Sanguinaria canadensis*. Oecologia 146: 32–37
Ramirez W (1970) Host specificity of fig wasps (Agaonidae). Evolution 24: 680–691
Rattenbury J (1979) Fruit-setting in *Sophora microphylla* Ait. NZ J Bot 17: 423–424
Rickson FR (1977) Progressive loss of ant-related traits of *Cecropia peltata* on selected Caribbean islands. Am J Bot 64: 585–592
Vovides AP (1991) Insect symbionts of some Mexican cycads in their natural habitat. Biotropica 23: 102–104
Walls GY (1978) The influence of the tuatara on fairy prion breeding on Stephens Island, Cook Strait. NZ J Ecology 1: 91–98
Ware AB, Compton ST (1992) Breakdown of pollinator specificity in an African fig tree. Biotropica 24: 544–549
Watt KEF, Molloy LF, Varshney CK, Weeks D, Wirosardjono S (1977) The unsteady state. East-West Center, University of Hawaii, Honolulu
Wiebes JT (1979) Co-evolution of figs and their insect pollinators. Annu Rev Ecol Syst 10: 1–12
Wilson EO (1987) The arboreal ant fauna of Peruvian Amazon forest: a first assessment. Biotropica 19: 245–251
Wolin CL (1985) The population dynamics of mutualistic systems. In: Boucher DH (ed) The biology of mutualism. Croom Helm, London, pp 248–269

13 Biological Diversity and Disturbance Regimes in Island Ecosystems

D. MUELLER-DOMBOIS

13.1 Introduction

The role of disturbance in natural communities and ecosystems has received a great deal of attention in the literature. For recent reviews see Cairns (1980); Burgess and Sharpe (1981); West et al. (1981); Pickett and White (1985); Remmert (1991); and Goldammer (1992). The concept of succession is based on disturbance as the initiating cause of a chronosequence in community and vegetation development. The successional changes following a disturbance are often measured by changes in biological diversity.

Some disturbances are so obvious, for example, a recent fire, a lava flow, or the scarification of the forest floor by pig activity, that the factual evidence or causes of such disturbances need no analysis. In such cases, usually the recovery process becomes an ecological research task. In many other situations, however, the causes of the disturbances are hidden. Instead, we may observe the effects of disturbances in the vegetation and associated biota. In such cases, for example, the decline or dieback of a forest canopy, the analysis of causal factors becomes another research challenge. In both cases, the focus is on vegetation dynamics, whose functioning is intimately associated with biological diversity and the disturbance regime. However, the question is, in what way?

In this context I will explore three questions:

1. Are climatic disturbance regimes in islands among the determinants of biological diversity?
2. Is biological diversity a factor in the recovery process following a disturbance?
3. Can biological diversity be a factor in the breakdown process of an established community?

I define disturbance regime as a repeating pattern of interferences in ecosystem development. Since long-term ecological studies in oceanic islands are almost non-existent, it is difficult to make definitive statements about the interrelationship of biological diversity and disturbance regimes. Moreover, disturbance events are hard to predict in a spatially or temporally precise manner. However, a conceptual framework can be put together from inferences based on limited data.

Department of Botany, University of Hawaii at Manoa, 3190 Maile Way, Honolulu, Hawaii 96822-2279, USA

Ecological Studies, Vol. 115
Vitousek et al. (eds) Islands

With these limitations in mind, I will discuss the topic from three different perspectives that relate to the stated questions.

13.2 Disturbance Regimes and Biodiversity Patterns Across the Pacific Islands

13.2.1 Disturbance Regimes

Disturbance regimes can be broadly classified into climatic, volcanic, biotic, and other categories, such as mechanical, physiological, and nutritional. The latter three are interrelated and they can also be used as subcategories of the former, and the list can be extended.

Across the many Pacific archipelagoes one can recognize at least two major types of climatic disturbances. These are hurricanes or typhoons (both terms refer to the same high-intensity storm patterns of tropical cyclones) and the ENSO events (E1 Niño-Southern Oscillations). These affect geographically different areas with somewhat predictably different frequencies and intensities.

According to Visher (1925) and Thomas (1965), there are three major cyclone (hurricane or typhoon) zones in the Pacific area. In the NE Pacific, a zone is restricted to the coast of Mexico; it does not extent to the Hawaiian Islands. In the NW Pacific, a zone includes Guam and the Marianas as well as Taiwan and the Philippines, but not the Marshalls and Central Pacific Islands. In the SW Pacific, the third high-frequency and high-intensity cyclone zone includes Western Polynesia with Tonga, Samoa, and the Cook Islands, and Melanesia with Fiji, Vanuatu, the Santa Cruz Islands, and New Caledonia, but less so the Solomons, Bougainville, or New Guinea. Thus, most of Eastern Polynesia is also out of the tropical cyclone zone.

Strong ENSO events which occur with a frequency of about ten per century (Quinn et al. 1987; Mueller-Dombois 1992a) reverse the atmospheric pressure system across the tropical Pacific. Normally, a low-pressure system prevails over the SW Pacific with the highest sea-level rainfall of about 5000 mm/year centering on the Santa Cruz Islands between Vanuatu and the Solomons. During a major ENSO event, the low pressure system jumps to the eastern Pacific in conjunction with a massive equatorial flow of warm surface water. During such times, excessive rainfall with enormous flash floods can be experienced in the normally dry Galápagos Islands (Robinson and del Pino 1985), while an extended drought can be expected in the normally wet SW Pacific to Indonesian areas (Goldammer 1990). Ramage (1986) has pointed to an eastward shift of tropical cyclone activity during the major ENSO events.

While the hurricanes are ordinarily known to cause extensive mechanical damage in the SW Pacific Island forests (Whitmore 1981), major ENSO events can be expected to cause interruptions of the normal growth patterns in island forests through physiological shocks with associated biological changes

(Kastdalen 1982; Lawesson 1988). They may also predispose normally wet rainforests to fire (Goldammer 1990).

13.2.2 Biodiversity Patterns

Biological diversity can be differently interpreted. Most commonly recognized are taxonomic richness and genetic diversity. Taxonomic richness has been shown by number of indigenous seed plant genera to be related primarily to size of archipelago area by Ash (1992). If frequent hurricanes had significant effects on lowering taxonomic richness, Vanuatu, in the high-frequency hurricane zone of the SW Pacific, should display a decreased taxonomic richness as compared to the Society Islands. If frequent hurricanes encouraged taxonomic diversity, the opposite trend should be expected. Vanuatu has a higher taxonomic richness in accordance with its greater land area. The 15 Pacific island groups analyzed by Ash (1992) fall into a remarkably straight line of number of genera over island size.

Another determinant of taxonomic richness is distance to source area. A richness-distance relationship has been shown by Woodroffe (1987) for mangrove species (see also Ash 1992). The number of mangrove species decreases from 33 in New Guinea to 20 in the Solomon Islands, 10 in Vanuatu and New Caledonia, 7 in Fiji, 4 in Samoa, and 1 each in the Marshall and Mariana Islands. This trend of species loss is functionally related to distance from source area. Any correlation to climatic disturbance regime would be only spurious.

These two examples may be sufficient to point out that taxonomic richness, which is an important element of biological diversity, responds functionally to island size and distance rather than disturbance regime. Other parameters functionally related to taxonomic diversity in islands are topography and habitat diversity as well as geological and biogeographical history.

Habitat or landscape diversity and geological history may override those of island size and distance as determinants of taxonomic richness. This can be shown with New Caledonia. This island is a true surface remnant of Gondwana that, like New Zealand, broke off the eastern edge of the former southern continent about 80 million years ago by sea-floor spreading (Raven and Axelrod 1974). New Caledonia carried the Gondwana flora some distance into the Pacific, when between 38 and 28 million years ago another geological disturbance emplaced a huge mass of ultramafic material onto New Caledonia (Holloway 1979). It covered more than half of New Caledonia's original Gondwanic surface and must have caused considerable displacement, if not extinction, of some of its Gondwanic flora. Today, ultrabasic rock material covers one third of New Caledonia's surface (Jaffré 1980) with a unique flora mostly reassembled from the former Gondwanic elements through evolutionary adaptation to this rather toxic soil substrate. Morat et al. (1984) report the native flora to consist of 3256 vascular plant species with 76% endemism. In contrast, New Zealand, which is seven times larger, has a much smaller flora of 2300 vascular species with 85% endemism (Wardle 1991).

The differences can be attributed to the disturbance history which includes, in addition to the catastrophic nutritional disturbance caused by the emplacement of ultramafic substrate also volcanism, mountain building, and erosion, and climate change with associated changes in sea-water level. The latter brought about changes in former land bridge connections which aided in the invasion of more recent elements of the Indo-Malesian flora. A principal outcome was "species packing", which further explains the unusual biological diversity of New Caledonia (Morat et al. 1984). However, these historic disturbance events that explain today's differences in taxonomic richness are beyond the consideration of this chapter. Instead, disturbance regimes, which are here defined as repeating patterns of interferences in ecosystem development, should be seen as relating to ecological time, which may be considered as the time frame since the last glaciation, about 10 000 years ago.

13.2.3 The Concept of Biological Diversity

The concept of biological diversity is not restricted to taxonomic richness or genetic diversity. From an ecological viewpoint the concept includes the quantitative relationships among the taxonomic organism groups of an area or ecosystem. This is included in all diversity indices. Thus, two communities with the same number of plant species can have different diversity indices. Another diversity measure is that by life-form and function, which is ecologically more meaningful. A forest with trees, shrubs, ferns, other herbaceous taxa, and bryophytes and lichens is more diverse than a species-rich grassland, even if both yielded the same diversity indices. Oceanic islands are known to lack some significant biotic elements or functional life-form groups that are present on nearby continents and/or continental islands. The absence of terrestrial mammals is a well-known example. In the vegetation of the Pacific Islands, tall-growing conifers such as *Araucaria* and *Agathis* species disappear when one traverses from North Australia via the continental Melanesian Islands to the oceanic islands east of the Andesite Line.

South Pacific island forests with emergent conifers, which represent a significant structural and floristic diversity element, also impose a functional difference on the mixed-species hardwood forest that typically forms the lower canopy (Enright 1993).

Thus, in answer to the principal question posed for this section, taxonomic richness does not appear to be functionally related to the broader natural disturbance regimes operative in the Pacific region. Quantitatively modified richness in terms of diversity indices may show some relationships, but that would be difficult to demonstrate. Certainly, human-induced changes in disturbance regimes, such as the use of fire, which promotes the spread of pyrophytes, reduces species richness, and also life-form diversity. In contrast to mere richness of species/area or taxonomic richness, species diversity in ecological life-form groups may be influenced by natural disturbance regimes. What can be expected (in agreement with Denslow 1985) is that a greater species richness is found among those life-form groups that display ecological properties or strategies that are well adapted to the respective disturbance regime. This will be further explained below.

13.3 Disturbance Regimes and Biodiversity as Factors in Ecosystem Development

13.3.1 Factors in Ecosystem and Vegetation Development

Ecosystem development has been effectively defined by the five state factors of Jenny (1980), originally used by Jenny (1941) to explain soil formation, and later also by Major (1951) to explain community formation. These five factors can be stated in form of an integral function (f) as:

ecosystem = f (cl, pm, r, o, t),

where cl = climate, pm = parent material (including soil), r = relief or topography, o = organisms, and t = time.

The environmental component or habitat of an ecosystem is here defined by the state factors climate (cl), soil and parent material (pm), and topography or relief (r).

For the purpose of understanding vegetation development, the focus is on the living plants, the primary producers, of an ecosystem. To explain vegetation or plant community development, the organism component needs to be subdivided into at least three factors, which are:

fl = the flora of a phytogeographic province,
ac = accessibility of the plant species to the site of an ecosystem,
el = the ecological life-form attributes of the plant species.

Vegetation formation can then also be explained as an integral function of five state factors (Mueller-Dombois and Ellenberg 1974):

vegetation = f (fl, ac, el, h, t).

Here h = habitat is defined by the three state factors climate (cl), soil and parent material (pm), and topographic position or relief (r), which are not listed in the above equation. Instead, the biological component is emphasized by three factors (fl, ac, el). Here fl = flora has an analogous function as source material for vegetation development as has pm = parent material for soil development.

In this concept of vegetation or plant community formation, biological diversity is limited by:

1. whatever species are or become available in the regional flora (fl),
2. by their dispersal capacities or the accessibility factor (ac), and
3. by their ecological life-form properties, life-history attributes, or survival and growth strategies (el).

Again, disturbance is not explicity included as a separate factor in this concept of community formation, but disturbance certainly plays an important role in community development.

The classic use of the term "development" implies a progressive buildup of the organismic or biological component (o = fl, ac, el) of an ecosystem towards an increased state of organization. This increased state of organization may never come into an equilibrium with the environmental habitat factors (h = cl, pm, r) due to disturbances that counteract this developmental trend.

However, if not totally destructive, disturbances that can be recognized to form a regime will become effective as periodically recurring stresses or transient stress factors. As such they must be viewed as long-persisting but oscillatory environmental or ecological forcing factors whose perturbing influences on shaping the biota may be equally as strong and effective as those of the more constantly persisting environmental forcing factors, here recognized as climate (cl), soil (pm), and topographic position (r). In order to become effective as a disturbance regime in ecosystem development, these oscillatory stresses must recur over a long time span, lasting at least several centuries.

As such they may act in two ways:

1. by selecting from the available regional flora pre-adapted plants that can cope successfully with the periodically recurring stresses, and
2. by molding plants from the existing flora that are moldable in terms of phenotypic plasticity or genetic adaptability, or both.

Thus, recurring perturbations will yield an adaptive response among the organisms of an ecosystem, and it may be reasonable to assume that the richer the original flora or taxonomic diversity of an area or island, the greater will be the number of species that can evolve adaptive traits or strategies. This, to me, is a most compelling argument for species preservation, particularly in view of a changing climate.

The question, whether biological diversity is a factor in the recovery of a disturbed ecosystem must be viewed again from the perspective of species diversity within life-form types, i.e., ecological species diversity rather than taxonomic species diversity. Then case examples can be suggested.

13.3.2 Case Examples

1. In the humid montane zone of Sri Lanka, the dominant vegetation is a species-rich montane rainforest (Werner 1988). Within the matrix of this rainforest are large patches of open grassland, which are spread across an undulating plateau near 2100 m elevation. The grassland has been in existence for centuries, and thus had posed the question of a climatic (frost-induced) origin (Mueller-Dombois and Perera 1971). Anthropogenic fires are now occurring almost annually in some topographic positions, while others have been kept unburned for a decade or two (Balakrishnan 1977). The boundary between forest and grassland is typically abrupt. The grassland contains only one tree species that has survived in the lower-frequency fire regime. This is *Rhododendron zeylanicum*, a tree with woody underground tubers. None of the

other tree species is able to reinvade the grassland; but natural restoration could probably be achieved by allowing the full recovery of the persisting *Rhododendron* population through fire protection. After recovery of this fire-tolerant pioneer population, the vigor of the grassland would be reduced by shading, thereby providing a re-invasion chance for other montane tree species from the surrounding forest.

2. A large-area clear-cutting of lowland tropical rain forest in Hawaii in 1983/84 resulted in the dominant recovery with herbaceous weed species, alien to Hawai'i. There was only one native sedge (*Machaerina angustifolia*), a native shrub (*Pipturus albidus*), and a native tree (*Metrosideros polymorpha*) among the indigenous colonizers that displayed a measurable abundance value. This represents an extremely low native diversity among a high alien species diversity of mostly herbs, ferns, grasses and shrubs. After 5 years, the recovering vegetation began to include a new group of ecological plant types, namely fast-growing secondary tropical tree species, such as *Paraserianthes* (*Albizia*) *falcataria* (*Fabaceae*), *Melochia umbellata* (Sterculiaceae), and *Trema orientalis* (Ulmaceae). In the recovery dynamics of this Hawaiian cut-over area, we now observe a totally new life-form group or ecologically different functional species group, a plant type that never developed any significant species richness among the natives.

   We can conclude from this that increased species diversity can very well speed up the recovery process after a disturbance; but again, this cannot be said without a qualifying statement, namely that the increased rate of recovery here is due to a new ecological group of fast-growing tree species that did not evolve in the Hawaiian environment. Of course, this aspect of diversity increase is not viewed as a desirable aspect of species diversity from the viewpoint of biological conservation.

3. Alien species invasion, considered to be a threat to native diversity, does not always lead to the extinction of native species. Much more commonly, it leads to species packing with somewhat predictable consequences for a changed dynamics of the vegetation.

In remote islands, such as the Hawaiian Islands, the Galápagos, Norfolk Island, and others, the addition of alien tree species will often add to the arrival of new ecological types, pre-adapted for a certain ecological role.

We know from succession studies of temperate forests that different tree species assume dominant roles along a chronosequence. Often pine stands, after fire, are followed or invaded by hardwoods, which in turn are invaded by spruces and firs just about when the pioneering pines are dropping out. Such a sequence represents a "chronosequential polyculture" on account of diversity changes with time, which occur on the same site. Instead, in islands one can often find "chronosequential monocultures" on account of the fact that the same species replaces itself as the dominant occupant on the same site. Thus, canopy species diversity may remain constantly low over long periods of time involving hundreds of generations. Such "direct" or auto-successions are known elsewhere from more extreme or harsh habitats, such as the subalpine fir forest in the Appalachian mountains (Sprugel 1976).

In contrast, lowland tropical rainforests on continents and continental islands, typically show a small-area successional pattern (gap dynamics) involving single tree falls as disturbance effects. The small-area disturbance patterns are typical for forests with species-rich canopies. From the viewpoint of successional diversity, such forests can be said to display a dynamics of "chronosequential gap formation". Species packing in island forests appears to promote a shift in vegetation dynamics from chronosequential monoculture to polyculture and gap rotation (Mueller-Dombois 1992b).

The question whether biological diversity can be regarded as a factor in the recovery process can certainly be answered with yes.

13.4 Disturbance Regime and Stand Demography

13.4.1 Disturbance as a Multivariate Regime

A disturbance regime can be determined by several variables. They include four scale parameters, namely, intensity, frequency, regularity, and area with regard to size and distribution. For a more complete analysis one also needs to consider the chance of return of a disturbance to the same area, and various forms of interaction among the disturbance parameters and the biological component of an ecosystem.

This set of variables or scale parameters coincides closely with those emphasized and defined as determinants by White and Pickett (1985). However, it would be incomplete to build a predictive model from these variables without the biological component which will vary in its predisposition depending on community development.

13.4.2 Developmental Stages in Stand Demography

Community development can be described as an interaction of the demographies of the participating populations. In a low-diversity community, the demography of a dominant or leading species may influence the whole community. In such cases we can characterize community development by stand demography. Oliver (1981) emphasized four demographic stages in stand development:

1. Stand reinitiation.
2. Stem exclusion.
3. Understory reinitiation.
4. Old growth.

In Oliver's concept of stand demography an intensive (or catastrophic) disturbance is implied prior to "stand reinitiation" of his chronosequence. This is clear because not only does the series start with a totally new stand but the undergrowth also must have been either wiped out or severely damaged prior to stand reestablishment. Thus, Oliver's sequence of stages appears adequate as a simplified pattern of primary succession.

However, in the course of stand demography, old growth is followed by mortality and breakdown, and breakdown is caused by disturbances.

What are these disturbances? The pathological viewpoint will implicate diseases and the ecological viewpoint old age? Both viewpoints may be correct, but neither viewpoint is an explanation of the disturbance regime.

13.4.3 Interaction Between Scale Variables and the Biological Community

For this we can use Oliver's (1981) stand establishment sequence; but we need to consider a 5th stage, the breakdown stage for answering the third question.

1. If *stand reinitiation* after a catastrophic disturbance is dominated by an even-aged tree population we can speak of *cohort stand* establishment. The alternative is an *all-aged stand*, and biodiversity enters into this from the start.
2. The *stem exclusion* stage may result in the elimination of later arrivals of the same species, reinforcing the even-aged character of a leading cohort population.
3. *Understory reinitiation* may often be associated already with the stand reinitiation stage. This is certainly so where one deals with a pattern of initial floristic composition (as suggested by Egler 1954). The alternative may be a pattern of relay floristics (see McIntosh 1980), which was apparently considered more characteristic by Oliver (1981).
4. The *old growth* stage may be reached only in a relatively disturbance-free environment, but it may also follow a low-intensity disturbance, which initiates the breakdown of the majority of the mature individuals of a canopy cohort. Such a breakdown can be manifested in a mass dieback of the less vigorous members of the leading cohort. The remaining old-growth trees may be the genetically superior survivors.
5. The *breakdown* stage. In forest systems with extensive cohort stands, such as found in fire-regulated pine forests, or also in different island and mountain forests, there may be no real old-growth stage. A low-intensity disturbance can interact with the decreased vigor state of the senescing individuals of an aging cohort resulting in a large-area forest dieback. This has been further explained by the cohort senescence theory (Mueller-Dombois 1983; 1986; 1988; 1992c; 1993).

13.5 Tree Mortality Patterns as Mediated by Biodiversity and Disturbance Regime

How does biodiversity and disturbance regime relate to the tree mortality patterns in island ecosystems? The answer to this is one of the most striking examples on how island studies can contribute to understanding of biodiversity and ecosystem relationships.

In natural ecosystems we always encounter some dead trees. When we find these to be widely scattered in an otherwise green canopy matrix, we consider the forest healthy and explain the scattered dead trees as initiators of canopy gaps, i.e., as causes sensu Pickett and White (1985). When we encounter instead many dead standing trees, we suspect either a past fire or an insect pest, or we consider the forest unhealthy and affected by either a biotic disease such as a fungus or virus or by an abiotic stress such as air pollution. All of these suspicions are worthy as research hypotheses.

However, research in different Pacific islands, such as Hawai'i (Mueller-Dombois 1993; Jacobi 1993; Gerrish 1993; and Jeltsch and Wissel 1993), the Galápagos (Lawesson 1988; Itow and Mueller-Dombois 1988), New Zealand (Stewart and Veblen 1983; Wardle and Allen 1983; Ogden 1988; Ogden et al. 1993), Papua New Guinea (Arentz 1983; Ash 1988; Enright 1988), and Japan (Kohyama 1988) has revealed the need to consider yet another major factor complex in addition to disease and/or abiotic stress to explain canopy dieback or mass mortality at the stand or landscape level.

This additional factor complex is stand demography in a community context as mediated by biological diversity and the disturbance regime. These two elements are part of the cohort senescence theory. For more recent interpretations see Mueller-Dombois (1992c, 1993).

13.6 Summary and Conclusion

I defined disturbance regime broadly as a repeating pattern of interferences or upsets in ecosystem and community development. On the basis of a few examples of climatic disturbance regimes in the Pacific basin, I concluded that taxonomic richness is not related. However, historical disturbances can substantially contribute to taxonomic richness as shown by the example of New Caledonia. Such historical disturbances are isolated, nonsystematic events that cannot be considered disturbance regimes.

I defined biological diversity in a fourfold sense as referring to taxonomic or species richness of an area, its dominance relationships of species in ecologically functional life-form groups, as well as population age structure and genetic variability within the species of a community. I then discussed ecosystem and vegetation development and concluded that the species assemblage of plant communities is determined by three environmental factors (cl, pm, r) and by three biological factors (fl, ac, el) and also by the biome-specific disturbance regime.

The concept of disturbance regime was then analyzed in more detail as consisting of four scale parameters: intensity, frequency, regularity, and area in terms of size and distribution. These parameters interact in a multivariate manner in the development and breakdown processes of the biological community of an ecosystem. On the basis of several island examples, I concluded that species diversity in ecologically functional species groups is critical in both the recovery and dieback processes.

A modeling effort for different forms and spatial configurations of forest stand mortality or dieback have recently been attempted by Jeltsch and Wissel (1993). These are based on the cohort senescence theory, which includes the above outlined disturbance regime and its interaction with the biological diversity of a canopy stand in the sense of species richness, dominance relations, age structures, and genetic variability. Thus biological diversity can be considered as functionally involved in ecosystem processes. Island studies have been essential in the establishment of this model.

Acknowledgments. The underlying work was supported by NSF Grants BSR-8718994 and BSR-891826 to the author. I thank my wife, Annette Mueller-Dombois, for continued support and active help in doing all the word processing.

References

Ash J (1988) *Nothofagus* (Fagaceae) forest on Mt. Giluwe, New Guinea. N Z J Bot 26: 245–258

Ash J (1992) Vegetation ecology of Fiji: past, present and future perspectives. Pac Sci 46: 111–127

Arentz F (1983) *Nothofagus* dieback on Mt. Giluwe, Papua New Guinea. Pac Sci 37: 453–458

Balakrishnan N (1977) Succession on abandoned fields in the Black Patana, a montane grassland in Sri Lanka (Ceylon); MSc Thesis, Botany Dept University of Hawaii, Honolulu

Burgess RL, Sharpe DM (1981) Forest island dynamics in man-dominated landscapes. Ecological Studies, Vol. 41. Springer, Berlin Heidelberg New York, 310 pp

Cairns J Jr (ed) (1980) The recovery process in damaged ecosystems. Ann Arbor Science Publishers, Ann Arbor, MI 167 pp

Denslow JS (1985) Disturbance-mediated co-existence of species. In: Pickett STA, White PS (eds) The ecology of natural disturbance and patch dynamics. Academic Press, Orlando, pp 307–323

Egler FE (1954) Vegetation science concepts I. Initial floristic composition, a factor in old-field vegetation development. Vegetatio 4: 412–417

Enright NJ (1993) Group death of *Araucaria hunsteinii* K. Schumm (Klinkii pine) in a New Guinea rainforest. In: Huettl RF, Mueller-Dombois D (eds) Forest decline in the Atlantic and Pacific regions. Springer, Berlin Heidelberg New York, pp 321–331

Gerrish G (1993) Using a life-history carbon balance model for forest-decline research. In: Huettl RF, Mueller-Dombois D (eds) Forest decline in the Atlantic and Pacific regions. Springer, Berlin Heidelberg New York, pp 243–250

Goldammer JG (ed) (1990) Fire in the tropical biota. Ecological Studies, Vol. 84. Springer, Berlin Heidelberg New York, 497 pp

Goldammer JG (ed) (1992) Tropical forests in transition: ecology of natural and anthropogenic disturbance processes. Birkhäuser, Basel, 270 pp

Holloway JD (1979) A survey of Lepidoptera. Biogeography and ecology of New Caledonia. Dr. W Junk, The Hague, 588 pp

Itow S, Mueller-Dombois D (1988) Population structure, stand-level dieback and recovery of *Scalesia pedunculata* forests in the Galapagos Islands. Ecol Res 3: 333–339

Jacobi JD (1993) Distribution and dynamics of *Metrosideros* dieback on the island of Hawaii: implications for management programs. In: Huettl RF, Mueller-Dombois D (eds) Forest decline in the Atlantic and Pacific regions. Springer, Berlin Heidelberg New York, pp 236–242

Jaffré P (1980) Végétation des roches ultrabasiques en Nouvelle Calédonie. Étude écologique du peuplement vegetal des sols dérivés de roches ultrabasiques en Nouvelle Calédonie. ORSTOM, Paris, 273 pp
Jeltsch F, Wissel C (1993) Modelling factors which may cause stand-level dieback in forests. In: Huettl RF, Mueller-Dombois D (eds) Forest decline in the Atlantic and Pacific regions. Springer, Berlin Heidelberg New York, pp 251–260
Jenny H (1941) Factors of soil formation. McGraw-Hill, New York, 281 pp
Jenny H (1980) The soil resource: origin and behavior. Springer, Berlin Heidelberg New York, 377 pp
Kastdalen A (1982) Changes in the biology of Santa Cruz – 1935–1965. Noticias Galapagos 35: 7–12
Kohyama T (1988) Etiology of "Shimagare" dieback and regeneration in subalpine *Abies* forests of Japan. GeoJournal 17: 201–208
Lawesson JE (1988) Stand-level dieback and regeneration of forests in the Galápagos Islands. Vegetatio 77: 87–93
Major J (1951) A functional, factorial approach to plant ecology. Ecology 32: 392–412
McIntosh RP (1980) The relationship between succession and the recovery process in ecosystems. In: Cairns J Jr (ed) The recovery process in damaged ecosystems. Ann Arbor Science Publishers, Ann Arbor, MI, pp 11–62
Morat P, Veillon J-M, Mackee HS (1984) Floristic relationships of New Caledonian rain forest phanerogams. In: Radovsky FJ, Raven PH, Sohmer HS (eds) Biogeography of the tropical Pacific. Assoc Syst Collections B P Bishop Museum, Honolulu, pp 71–128
Mueller-Dombios D (1983) Population death in Hawaiian plant communities: a new causal theory and its successional significance. Tuexenia (Festschrift H Ellenberg) 3: 117–130
Mueller-Dombois D (1985) Ohi a dieback in Hawaii: 1984 synthesis and evaluation. Pac Sci 39: 150–170
Mueller-Dombois D (1986) Perspectives for an etiology of stand-level dieback. Annu Rev Ecol Syst 17: 221–243
Mueller-Dombois D (1988) Canopy dieback and ecosystem processes in the Pacific area. Congress Plenary Lecture. In: Greuter W, Zimmer B (eds) Proc XIV Int Bot Congress, Berlin, Koeltz, Koenigstein/Taunus, pp 445–465
Mueller-Dombois D (1991) The mosaic theory and the spatial dynamics of natural dieback and regeneration in Pacific forests. In: Remmert H (ed) The mosaic-cycle concept of ecosystems. Ecological Studies, vol. 85. Springer, Berlin Heidelberg New York, pp 46–60
Mueller-Dombois D (1992a) Potential effects of the increase in carbon dioxide and climate change on the dynamics of vegetation. Water Air Soil Pollut 64: 61–79
Mueller-Dombois D (1992b) Distributional dynamics in the Hawaiian vegetation. Pac Sci 46: 221–231
Mueller-Dombois D (1992c) A natural dieback theory, cohort senescence as an alternative to the decline disease theory. In: Manion PD, Lachance P (eds) Forest decline concepts. APS Press. The Phytopath Soc, St Paul, Minnesota pp 26–37
Mueller-Dombois D (1993) Biotic impoverishment and climate change: global causes of forest decline? In: Huettl RF, Mueller-Dombois D (eds) Forest decline in the Atlantic and Pacific regions. Springer, Berlin Heidelberg New York, pp 338–348
Mueller-Dombois D, Ellenberg H (1974) Aims and methods of vegetation ecology. John Wiley, New York, 547 pp
Mueller-Dombois D, Perera M (1971) Ecological differentiation and soil fungal distribution in the montane grasslands of Ceylon. Ceylon J Sci Biol Sci 9: 1–41
Ogden J (1988) forest dynamics and stand-level dieback in New Zealand's *Nothofagus* forests. GeoJournal 17: 225–230
Ogden J, Lusk ChH, Steel MG (1993) Episodic mortality, forest decline and diversity in a dynamic landscape: Tongariro National Park, New Zealand. In: Huettl RF, Mueller-Dombois D (eds) Forest decline in the Atlantic and Pacific regions. Springer, Berlin Heidelberg New York, pp 261–274
Oliver CD (1981) Forest development in North America following major disturbances. For Ecol Manage 3: 158–168

Pickett STA, White PS (1985) The ecology of natural disturbance and patch dynamics. Academic Press, New York, 427 pp
Quinn WH, Neal VT, de Mayolo SEA (1987) E1 Niño occurrences over the past four and a half centuries. J Geo Phys Res 92: 14, 449–14, 461
Ramage CS (1986) E1 Niño. Sci Am 254: 77–83
Raven PH, Axelrod DI (1974) Angiosperm biogeography and past continental movements. Ann Misso Bot Gard 61: 539–673
Remmert H (ed) (1991) The mosaic-cycle concept of ecosystems. Ecological Studies, Vol. 85. Springer, Berlin Heidelberg New York, 168 pp
Robinson G, del Pino EM (eds) (1985) E1 Niño en las Islas Galápagos: E1 evento de 1982–1983. Contrib No 388, Charles Darwin Foundation Quito, Eccador, 534 pp
Sprugel DG (1976) Dynamic structure of wave-regenerated *Abies balsamea* forests in the northeastern United States. J Ecol 64: 889–911
Stewart GH, Veblen TT (1983) Forest instability and tree mortality in Westland, New Zealand. Pac Sci 37: 427–431
Thomas WL Jr (1965) The variety of physical environments among the Pacific Islands. In: Fosberg FR (ed) Man's place in the island ecosystem. Bishop Museum Press, Honolulu, pp 7–37
Visher SS (1925) Tropical cyclones of the Pacific. Bernice P Bishop Mus Bull 20: 1–163
Vitousek PM, Walker LR, Whiteaker LD, Mueller-Dombois D, Matson PA (1987) Biological invasion of *Myrica faya* alters ecosystem development in Hawaii. Science 138: 802–804
Wardle JA, Allen RB (1983) Dieback in New Zealand *Nothofagus* forests. Pac Sci 37: 397–404
Wardle P (1991) Vegetation of New Zealand. Cambridge University Press, Cambridge, 672 pp
Werner W (1988) Canopy dieback in the upper montane rain forests of Sri Lanka. Geo Journal 17: 245–248
West DC, Shugart HH, Botkin DB (1981) Forest succession: concepts and application. Springer, Berlin Heidelberg New York, 517 pp
White PS, Pickett STA (eds) (1985) Natural disturbance and patch dynamics: an introduction. In: Pickett STA, White PS (eds) The ecology of natural disturbance and patch dynamics. Academic Press Orlando pp 3–13
Whitmore TC (1974) Change with time and the role of cyclones in tropical rain forest on Kolombangara, Solomon Islands. Commonwealth For Inst Pap 46. Oxford Holywell Press, Oxford
Woodroffe CD (1987) Pacific island mangroves: distribution and environmental settings. Pac Sci 41: 166–185

14 Effects of Diversity on Productivity: Quantitative Distributions of Traits

J. H. FOWNES

14.1 Introduction

Biodiversity relates to ecosystem function on islands in two distinct ways. First, the isolation of some islands from larger land masses has led to unique patterns of biodiversity, caused by limited numbers of species introductions, absence of certain "functional groups" of species, limited refugia during past climate changes, and development of endemism and adaptive radiation within taxa. The importance of these biodiversity patterns to ecosystem function is often revealed by biological invasions having direct effects on ecosystem function, such as the bird-dispersed nitrogen-fixing shrub *Myrica faya* altering the N cycle in montane Hawaiian forests (Vitousek and Walker 1989). Island patterns of biodiversity also lead to what might be considered indirect consequences of invasions, such as the vulnerability of island plant species to relatively recent changes such as introduction of goats (Baker and Reeser 1972; Spatz and Mueller-Dombois 1973), and consequent ecosystem degradation from overgrazing. Second, although the effects of biodiversity per se on the function of a given ecosystem are not always unique to islands, islands may present unique opportunities for their investigation. For example, the dominant role of a single tree species, *Metrosideros polymorpha*, in Hawaiian forests has permitted much clearer identification of controls of ecosystem carbon and nitrogen accumulation than would be possible in more diverse plant communities (Vitousek 1995).

It is useful to define a functional group as those species having similar traits affecting ecosystem processes (e.g., productivity, nutrient cycling, or hydrology) (Körner 1993; Vitousek and Hooper 1993). However, grouping species discretely must be to some extent arbitrary, and species grouped according to distribution along one trait may be grouped differently according to another. We can also define a "keystone species" as one that is alone in a functional group, that is, without other species having similar traits (Bond 1993; Schulze and Mooney 1993). From this definition, the role of keystone species is determined more by a property of the community, that is, presence or absence of similar species, than by inherent species traits.

The recognition that identifying functional groups and keystone roles depends on similarity of traits among and within species carries with it the idea

Department of Agronomy and Soil Science, 1910 East-West Rd., University of Hawai'i at Manoa, Honolulu, Hawaii 96822, USA

Ecological Studies, Vol. 115
Vitousek et al. (eds) Islands

that in most cases there is continuous, not discrete, distribution of traits, at least potentially. Often, situations will be too complex to explain by single-variable traits, such as where two or more traits are tied together in a tradeoff relationship, or where the interaction of two or more functional groups of species determines ecosystem structure and function. It is commonly believed that island systems have fewer members in functional groups, may altogether lack certain functional groups found on continents, and have more "keystone roles" for native species as well as more keystone roles that could be occupied by invading species. Invasions of island ecosystems therefore present opportunities to document the impact of single species on ecosystem function, provided that the invaders differ sufficiently in physiological traits from natives, or if they alter trophic structure or disturbance regimes (Vitousek 1990). Because all these arguments rest on assigning some degree of similarity or difference to species, this chapter addresses what metrics of similarity we can use to quantify the effects of biodiversity on ecosystem function.

14.2 Simple Distributions of Traits

Many traits may be viewed as simple or one-dimensional continua, such as concentration of a chemical in leaf litter, light-saturated photosynthetic rate, minimum leaf water potential, maximum height or lifespan, fire tolerance, litter flammability, maximum nitrogen fixation rate, or maximum dispersal distance. A large difference in any one of these traits from other species present in a community can be sufficient for a single species to control ecosystem function. In this sense, the effects are not necessarily related to diversity per se: a species could play a regulating role by reducing resource availability to a low enough level that addition or removal of other species has no effect. For example, a species able to reduce soil solution N concentration to low levels would affect ecosystem N cycling, interspecies competition for N, and thus plant succession (Wedin and Tilman 1993), but this role may be relatively unaffected by how many other species are present. In the long term, however, higher species diversity has the effect of increasing the "choice" of species potentially able to fill a certain role. Over evolutionary time, populations on isolated islands diversify at the population, subspecies, or species levels and thereby fill previously vacant roles.

An example from Hawai'i is the myrtaceous tree *Metrosideros polymorpha*, which has apparently differentiated into ecologically distinct populations filling different successional roles. In a common garden study contrasting seedlings from populations with pubescent versus glabrous leaves, Stemmermann (1983) found that leaves from pubescent varieties had more negative osmotic potentials and lower bulk modulii of elasticity, implying that their success in colonizing young volcanic substrates was related to turgor maintenance under drought. She suggested that the morphological, physiological, and ecological differences were so great that more than one introduction of *Metrosideros* may have occurred. However, *Metrosideros* populations varying morphologically showed some altitudinal differentiation in isozyme variation, but were nevertheless very similar genetically (Aradhya et al. 1993). A comparison of isozyme variation in three morphologically recognized *Metrosideros* species showed that *M. tremuloides* was genetically indistinguishable from *M. polymorpha*, suggesting that species divergence was relatively recent (Aradhya et al. 1991).

Another interesting aspect of *Metrosideros* in Hawai'i is the tendency for apparently even-aged stands to senesce and regenerate nearly simultaneously, termed cohort regeneration (Mueller-Dombois 1992 a,b). Although the synchronizing and triggering factors regulating the cohort pattern remain only partially known, it is also likely that the pattern of cohort regeneration and senescence may have impacts on ecosystem function. Do pulses of root, leaf, and wood litter, productivity, and reduced evapotranspiration lead to long-term dynamics that are different from those predicted from average conditions? It seems likely, given that these processes have nonlinear and interacting controls, but no quantitative estimates have yet been made. A similar tendency was shown in simulations of montane *Eucalyptus* forest, but the oscillations damped out after approximately three generations (Shugart et al. 1981). The frequency and variance of ecosystem fluctuations would appear to be linked to the relatively low biodiversity of these montane tropical forests.

More information is available on the effects of *Myrica faya*, an actinorhizal nitrogen-fixing shrub that is filling a keystone role by invading N-limited, open-canopied montane forests in Hawai'i (Vitousek and Walker 1989). By fixing several times the preinvasion nitrogen input rates, and by producing litter that releases nitrogen to other organisms, *Myrica* is altering both the rate and pattern of ecosystem development at these sites. It is not clear why native nitrogen fixers (e.g., *Acacia koa, Sophora chrysophylla*) have not filled this role, but it has been suggested that their seeds are not well suited for rapid long-distance dispersal, unlike the bird-dispersed *Myrica* (Vitousek and Walker 1989). The many extinctions of Hawai'ian birds (James, Chap. 8, this vol.) may include ones that formerly dispersed seed of native nitrogen fixing trees. Whatever the reasons, the native vegetation currently lacks trees that both disperse and colonize and fix substantial atmospheric nitrogen.

14.3 Distributions of Two or More Traits

Whenever functional groups are recognized by two or more traits, it is important to determine if the traits interact or are somehow mutually constrained. It is simpler conceptually if they do not interact, because observations and experiments tend to be simpler and easier to interpret, but it is not always easy to determine whether interactions are really occurring. For example, it has long been known that nitrogen release from decomposing leaves and litter is affected by initial concentrations of nitrogen and of classes of carbon compounds (e.g., lignin, soluble polyphenols). Indices based on various ratios, one way of expressing interaction between factors, have correlated well with decomposition, such as lignin:N for temperate forest litter (Melillo et al. 1982), or with mineralization rates, such as soluble polyphenol:N for tropical leguminous green manure (Palm and Sanchez 1991; Oglesby and Fownes 1992), and (lignin + polyphenol):N for tropical legumes (Fox et al. 1990). However, using fresh leaves and senesced litter from a wide range of tropical legumes and non-legumes, Constantinides and Fownes (1994) found that these ratio indices derived their predictive power essentially from containing the reciprocal of N concentration and that there was no evidence for multiplicative effects (interactions) of N and carbon fractions; neither did the obvious functional groups (litter versus fresh leaves, N fixer versus non-fixer) affect N dynamics other than as reflected in the basic continua of concentrations of N (primarily), and soluble polyphenols (secondarily). In this case, within fairly broad constraints (minimum and maximum concentrations of N, polyphenols, lignin), the various traits were assorted independently in two or more dimensions, suggesting that patterns of biodiversity on islands would affect ecosystem N dynamics based on what combinations were available among successful colonizers. Nevertheless, other examples show that interactions among or tradeoffs between traits can be important in understanding diversity effects on ecosystem function.

14.4 Complementarity and Tradeoffs Among Traits

An important way that diversity of species or functional groups can increase primary productivity is when species use several limiting resources differently or are complementary in their use of one resource. More diverse assemblages can be more productive when functional groups are matched complementarily: for example, when N fixers are mixed with non-fixers in N-limited ecosystems (Trenbath 1976). The matching of complementary traits differs from random assembly of traits in either agricultural or natural ecosystems. Although it might be argued that complementary combinations would tend to be prevalent in natural ecosystems, the *Myrica faya* example shows that islands may not meet such an expectation based on the assumption of potentially unlimited species availability.

Another example of complementary resource use increasing productivity is when the canopy of a shade-intolerant species occurs above a shade-tolerant species. However, it is possible to postulate a neutral effect of mixing shade tolerance classes (evenly codominant) or even a dysfunctional mixture (shade-tolerants above intolerants) where productivity is less in the more diverse system. This example suggests that how species are mixed, in this case their canopy architecture, determines the outcome of ecosystem function in addition to the nonrandom selection of complementary combinations, both of which may be more important than species number per se.

Patterns of complementary resource use may be linked to tradeoffs in physiological function. Tradeoff patterns are useful and appealing because they reduce the dimensionality of the possible combinations of traits. By reducing all possible combinations of traits to a linked set, often embodying unavoidable physiological tradeoffs, this approach has been useful in understanding patterns of resource use, vegetation distribution and composition, and succession (Tilman 1993).

The basic form of a tradeoff pattern in an inverse relation between a species' performance under one constraint versus another (Fig. 14.1). The scatter of points for each species indicates the range of phenotypic plasticity. In this example, species B has two populations with differing ranges of phenotypic plasticity, suggesting genetic differentiation between them. The vacant area between species A and B represents a vacant niche, which, if occupied by a single species, could become a keystone role in this ecosystem.

A generally well-understood physiological tradeoff is that high rates of light-saturated photosynthesis demand a large investment in metabolic apparatus, leading to high respiration costs, and thus reduced photosynthesis in dim light compared with shade-adapted species (Boardman 1977). It is important, yet little studied, to compare phenotypic and genotypic variation within a species with the range observed among species, because this comparison strongly affects whether species are in fact differentiated along simple or tradeoff axes of traits. Furthermore, although the physiological tradeoff between sun and shade leaves is well known, additional factors may be necessary to explain the colonization and growth relationships of groups of species.

A good example of quantifying tradeoffs in resource use is the photosynthetic response of four Queensland rainforest species (Thompson et al. 1992a, b). From field observations, two species of *Argyrodendron* were considered shade-tolerant colonizers of small gaps, *Flindersia brayleyana* was characteristic of a wide variety of sites, and *Toona australis* was considered shade-intolerant because it characteristically colonizes large gaps with its wind-dispersed seeds and rapid growth rate. Seedlings of the four species were grown in pots at three levels of light (ranging from full sun to a low level equivalent to the average light level in the understory), and two inorganic N levels (approximating the elevated N availability in gap soils and the relatively low concentrations in undisturbed forest soils).

Plotting the gas exchange results (Thompson et al. 1992b) as a tradeoff diagram of respiration in the dark (less negative indicating better adjustment to

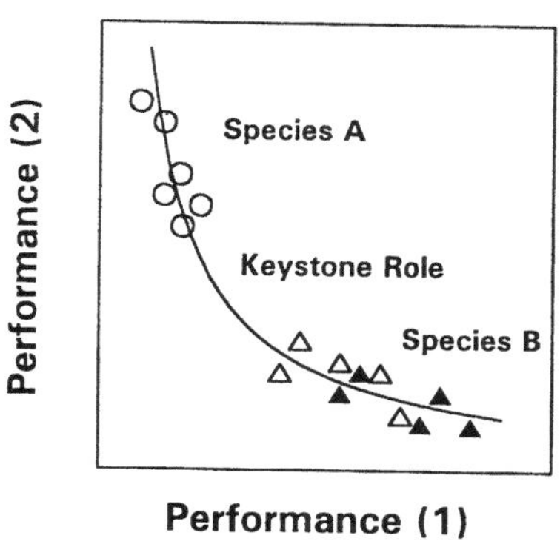

Fig. 14.1. Hypothetical tradeoff relationship, showing performance under one set of constraints being inversely related to performance under a second set of constraints. The *scatter of points* indicates phenotypic variation of species *A* and *B*, and the *filled* versus *open symbols* for species *B* represent two populations showing genetic differentiation. The distance between species *A* and *B* represents a potential keystone role to be filled by an invader

shade) versus light-saturated photosynthesis (positively related to performance in full sun), showed that taken together, the species did show a tradeoff in photosynthetic acclimation (Fig. 14.2). Three of the four species showed phenotypic adjustment to light and N, falling onto somewhat different spaces on the graph. Although *Toona* greatly increased maximum photosynthesis in the high-light, high-N treatment, its range of plasticity entirely spanned the range of the other species, as it also had the lowest (least negative) dark respiration in the low-light, low-N treatment. Contrary to expectation, *Toona* was able to adjust its photosynthetic function to the shade environment. Other traits, such as its low seed mass and less conservative carbohydrate storage pattern must be invoked to explain *Toona*'s general absence from the forest floor and dependence on gaps for recruitment (Thompson et al. 1992a). A similar conclusion was reached in a study of lowland Costa Rican forest, where a simple pioneer/shade tolerant dichotomy did not apply to relative growth rate or a tradeoff between survival in shade and capacity for rapid growth in full sunlight (Clark and Clark 1992). Although tradeoffs in physiological function are operative in these examples, they may not explain species' typical distribution (as much as we would like them to). Nevertheless, quantification of the traits affecting ecosystem function is a path to understanding effects of biodiversity in natural or managed ecosystems.

Harrington et al. (1989) studied patterns of growth and carbon gain in two native and two invading shrubs in southern Wisconsin. They found that all four species increased growth efficiency (biomass growth rate per unit leaf area) in open habitats, and increased leaf area ratio (leaf area per unit total biomass) in the understory. There appeared to be a tradeoff between a species' ability to respond to increased light in openings by increasing growth per unit leaf area, and its ability to shift its allocation away from wood to leaves in the understory (Fig. 14.3). In this case, the two invading species fell between the native species, suggesting that the two natives were dissimilar enough for a vacant role to be available, whereas the invading shrubs were more similar to each other. It is not likely that differences in sun-shade plasticity solely determined colonization success among species of understory shrubs, yet at least we have an unambiguous measure of species differences along two interacting trait axes.

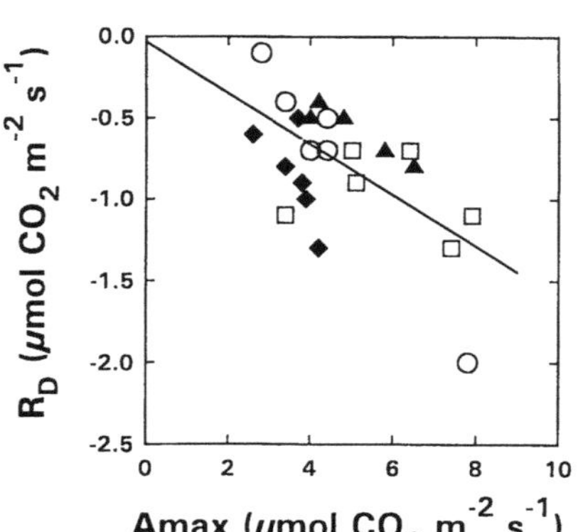

Fig. 14.2. The tradeoff between dark respiration (R_D) and maximum light-saturated photosynthesis (A_{max}) for four species of Queensland rainforest trees grown under various light and nitrogen regimes. Shade-tolerant species are *Argyrodendron* sp. (◆) and *Argyrodendron trifoliatum* (▲), and sun-tolerant species are *Flindersia brayleyana* (□) and *Toona australis* (○). The *line* is the regression for all points $R_D = -0.033 - 0.157\ A_{max}$, $r^2 = 0.36$, $P = 0.002$). (Data from Thompson et al. 1992b)

14.5 Biodiversity and the Usefulness of Production

In addition to affecting productivity per se, diversity in fruiting season, edibility, nutrition, disease resistance, and other properties strongly affects the usefulness to humans of ecosystem production, and high species and cultivar diversities characterize many indigenous agroecosystems of the tropics. In a randomized survey of the agroforestry land use system of the island of Pohnpei, Federated States of Micronesia, Raynor and Fownes (1991a) found 161 plant species in 10 ha of survey plots. Roughly two-thirds of these species were deliberately planted or cultivated, but many of the uncultivated species also had various uses such as food or traditional medicine. The major cultivated plant species also had very high diversity of cultivars: 28 breadfruit (*Artocarpus altilis*), 38 yam (*Dioscorea* spp.), 18 plantain (*Musa* spp.), and several each of banana (*Musa* spp.), coconut (*Cocos nucifera*), taro (*Colocasia* spp., *Alocasia* spp.), kava (*Piper methysticum*), and sugarcane (*Saccharum officinarum*) cultivars were found in the survey area. Based on literature review and farmer interviews, the breadfruit and yam cultivars found in the survey plots represent only about 20% of the cultivar names known to Pohnpeian farmers (Raynor and Fownes 1991a).

An example of the importance of quantitative traits in assessing the role of biodiversity is complementarity in fruiting among breadfruit cultivars (Fownes and Raynor 1993). Because breadfruit has a distinct fruiting season when ripe fruit is abundant, followed by periods of no available fruit, several of the most important breadfruit cultivars were monitored for 1 year. These cultivars showed distinctly offset fruiting periods (Fig. 14.4), implying that farmers who maintained higher cultivar diversity could benefit from an extended period of food availability, as well as diversity in flavor, storage potential, disease resistance, and other

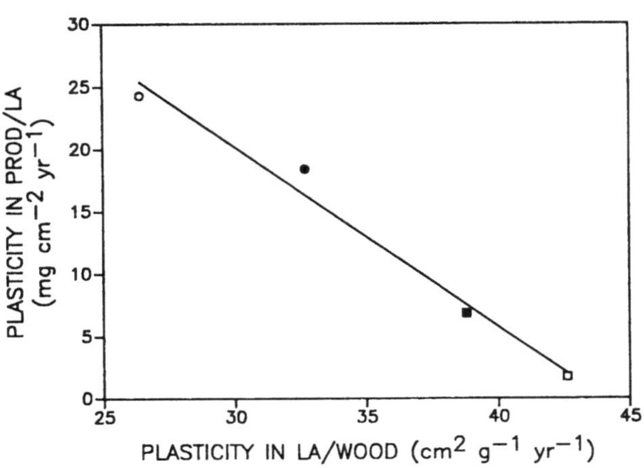

Fig. 14.3. Difference in production per unit leaf area between open and understory sites versus difference in leaf area: wood ratio between open and understory sites for the invading species *Rhamnus cathartica* (●) and *Lonicera* X *bella* (■), and the native species *Cornus racemosa* (○) and *Prunus serotina* (□) in southern Wisconsin. (Harrington et al. 1989)

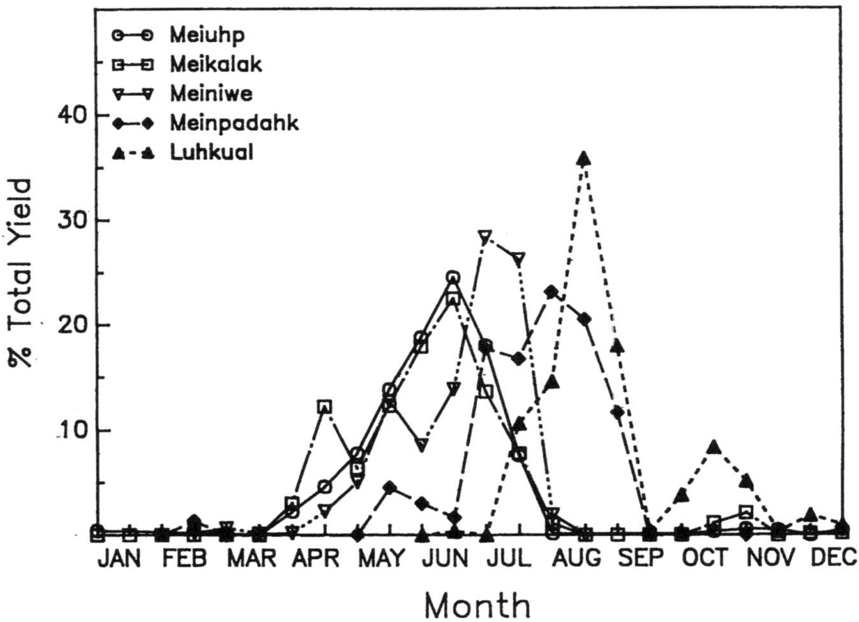

Fig. 14.4. Proportion of total annual breadfruit yield occurring in 2-week intervals for five traditional cultivars, Pohnpei Island. (Fownes and Raynor 1993)

traits (Fownes and Raynor 1993). It is not known what environmental cues controlled the variation in fruiting time, but farmers' lore agreed with the pattern found in the study year, suggesting that the pattern was repeatable. The mean harvestable yield of breadfruit in the agroforest landscape was relatively high for a low-input system (6.7 t ha^{-1} $year^{-1}$, fresh weight basis) (Fownes and Raynor 1993), especially given that breadfruit trees constituted only about 25% of the total number of canopy trees (those greater than 8 m height) (Raynor and Fownes 1991a). The complementarity of fruiting period among cultivars is probably more important for the utilization of production than for the total amount of production, although there may be an interaction with the buildup of spore populations and thus disease incidence.

Another dimension of the diversity of the indigenous Pohnpeian agroforestry system is variation in space and time. The landscape diversity of the island as a whole includes mangrove and upland forest as well as settled agroforest. Within a landholding in the agroforestry area, the typical pattern is concentric zones around the house compound, generally showing decreased management intensity as one moves farther away (Raynor and Fownes 1991b). The spatial pattern interacts with successional time, in that food crops and kava are planted around the household in young farms, then extended outward over time. The density of secondary successional species (e.g., *Hibiscus tiliaceus*) is determined in a dynamic balance between labor availability and invasion and growth rates. Important exceptions to this general radial pattern are plantain, grown near the house in all stages (because it is considered a "woman's crop"), and yams, which are widely dispersed, if not hidden, at all stages of farm development (Raynor and Fownes 1991b). These studies illustrate how biodiversity at several scales (cultivar, species, landscape, and succession) is harnessed by people to utilize ecosystem productivity.

14.6 Conclusions

In island ecosystems, native species play keystone roles, and invaders can fill vacant keystone roles, because of low diversity related to islands' relative isolation. On the other hand, the evolutionary diversification of *Metrosideros polymorpha*, and the rich cultivar diversity utilized in indigenous Pacific island land-use systems show that ecosystem-level function can be affected by biodiversity within a species. Concepts such as "keystone species" and "functional group" depend on the context of the rest of the community as well as distances along axes of physiological traits that are usually continuous. A single species can alter ecosystem function if it differs in one trait from other species present sufficiently to alter resource supply, yet patterns of community composition and change may be regulated by two or more sets of traits, acting either independently or interacting in complementary or tradeoff relationships. Both phenotypic plasticity and genetic variability in quantitative traits affecting ecosystem function must be assessed to reckon differences among species.

References

Aradhya KM, Mueller-Dombois D, Ranker TA (1991) Genetic evidence for recent and incipient speciation in the evolution of Hawaiian *Metrosideros* (Myrtaceae). Heredity 67: 129–138

Aradhya KM, Mueller-Dombois D, Ranker TA (1993) Genetic structure and differentiation in *Metrosideros polymorpha* (Myrtaceae) along altitudinal gradients in Maui, Hawaii. Genet Res Camb 61: 159–170

Baker JK, Reeser DW (1972) Goat management problems in Hawaii Volcanoes National Park: a history, analysis, and management plan. Natural Resources Rep No 2, National Park Service Publ NPS-112. US Dept of Interior, Washington, DC

Boardman NK (1977) Comparative photosynthesis of sun and shade plants. Annu Rev Plant Physiol 28: 355–377

Bond WJ (1993) Keystone species. In: Schulze E-D, Mooney HA (eds) Biodiversity and ecosystem function. Springer, Berlin Heidelberg New York, pp 238–253

Clark DA, Clark DB (1992) Life history diversity of canopy and emergent trees in a neotropical rain forest. Ecol Monogr 62: 315– 344

Constantinides M, Fownes JH (1994) Nitrogen mineralization from leaves and litter of tropical plants: relationship to nitrogen, lignin and soluble polyphenol concentrations. Soil Biol Biochem 26: 49–55

Fownes JH, Raynor WC (1993) Seasonality and yield of breadfruit cultivars in the indigenous agroforestry system of Pohnpei, Federated States of Micronesia. Trop Agric (Trinidad) 70: 103–109

Fox RH, Myers RJK, Vallis I (1990) The nitrogen mineralization rate of legume residues in soil as influenced by their polyphenol, lignin, and nitrogen contents. Plant Soil 129: 251–259

Harrington RA, Brown BJ, Reich PB, Fownes JH (1989) Ecophysiology of exotic and native shrubs in Southern Wisconsin. II. Annual growth and carbon gain. Oecologia 80: 368–373

Körner Ch (1993) Scaling from species to vegetation: The usefulness of functional groups. In: Schulze E-D, Mooney HA (eds) Biodiversity and ecosystem function. Springer, Berlin Heidelberg New York, pp 117–140

Melillo JM, Aber JD, Muratore JF (1982) Nitrogen and lignin control of hardwood leaf litter decomposition dynamics. Ecology 63: 621–626

Mueller-Dombois D (1992a) A global perspective on forest decline. Environ Toxicol Chem 11: 1069–1076

Mueller-Dombois D (1992b) Potential effects of the increase in carbon dioxide and climate change on the dynamics of vegetation. Water Air Soil Pollut 64: 61–79

Oglesby KA, Fownes JH (1992) Effects of chemical composition on nitrogen mineralization from green manures of seven tropical leguminous trees. Plant Soil 143: 127–132

Palm CA, Sanchez PA (1991) Nitrogen release from the leaves of some tropical legumes as affected by their lignin and polyphenolic contents. Soil Biol Biochem 23: 83–88

Raynor WC, Fownes JH (1991a) Indigenous agroforestry of Pohnpei. 1. Plant species and cultivars. Agrofor Syst 16: 139–157

Raynor WC, Fownes JH (1991b) Indigenous agroforestry of Pohnpei. 2. Spatial and successional vegetation patterns. Agrofor Syst 16: 159–165

Schulze E-D, Mooney HA (1993) Ecosystem function of biodiversity: A summary. In: Schulze E-D, Mooney HA (eds) Biodiversity and ecosystem function. Springer, Berlin Heidelberg New York, pp 497–510

Shugart HH, West DC, Emanuel WR (1981) Pattern and dynamics of forests: An application of simulation models. In: West DC, Shugart HH, Botkin DB (eds) Forest succession: concepts and application. Springer, Berlin New York, pp 74–94

Spatz G, Mueller-Dombois D (1973) The influence of feral goats on koa tree reproduction in Hawaii Volcanoes National Park. Ecology 54: 870–876

Stemmermann L (1983) Ecological studies of Hawaiian *Metrosideros* in a successional context. Pac Sci 37: 361–373

Thompson WA, Kriedemann PE, Craig IE (1992a) Photosynthetic response to light and nutrients in sun-tolerant and shade-tolerant rainforest trees. I. Growth, leaf anatomy and nutrient content. Aust J Plant Physiol 19: 1–18

Thompson WA, Huang L-K, Kriedemann PE (1992b) Photosynthetic response to light and nutrients in sun-tolerant and shade-tolerant rainforest trees. II. Leaf gas exchange and component processes of photosynthesis. Aust J Plant Physiol 19: 19–42

Tilman D (1993) Community diversity and succession: The roles of competition, dispersal, and habitat modification. In: Schulze E-D, Mooney HA (eds) Biodiversity and ecosystem function. Springer, Berlin Heidelberg New York, pp 327–344

Trenbath BR (1976) Plant interactions in mixed crop communities. In: Papendick RI, Sanchez PA, Triplett GB (eds) Multiple cropping. American Society of Agronomy, Madison, WI, pp 129–170

Vitousek PM (1990) Biological invasions and ecosystem processes: Towards an integration of population biology and ecosystem studies. Oikos 57: 7–13

Vitousek PM (1995) The Hawaiian Islands as a model system for ecosystem studies. Pac Sci (in press)

Vitousek PM, Hooper DU (1993) Biological diversity and terrestrial ecosystem biogeochemistry. In: Schulze E-D, Mooney HA (eds) Biodiversity and ecosystem function. Springer, Berlin Heidelberg New York, pp 3–14

Vitousek PM, Walker LR (1989) Biological invasion by *Myrica faya* in Hawai'i: Plant demography, nitrogen fixation, ecosystem effects. Ecol Monogr 59: 247–265

Wedin D, Tilman D (1993) Competition among grasses along a nitrogen gradient: initial conditions and mechanisms of competition. Ecol Monogr 63: 199–229

Section D
Conservation Implication

15 Insular Lessons for Global Biodiversity Conservation with Particular Reference to Alien Invasions

I.A.W. MACDONALD and J. COOPER

15.1 Introduction

The SCOPE Programme on the Ecosystem Function of Biodiversity is aimed at ascertaining whether it is necessary to prevent the predicted man-induced extinction of much of the Earth's current biodiversity (Ricklefs et al. 1990) on the grounds of securing the adequate functioning of global life-support systems. Given this background, it would appear inappropriate to focus on oceanic islands, as they constitute a minute fraction of the Earth's total land surface and, accordingly, are likely to make an infinitesimally small contribution to global fluxes in the biosphere. Even on simple biodiversity conservation grounds, a focus on islands within a global programme is of doubtful validity, as most biodiversity is located on the continents and in the oceans.

However, insular vertebrate faunas have given rise to the vast majority of historical extinctions (Honegger 1981; King 1985) and to a high proportion of the species currently thought to be endangered (31% of the 941 vertebrate species – excluding fishes – so listed in 1986 by the World Conservation Monitoring Centre: Macdonald et al. 1989). Given the recent, predominantly reactive, mode of conservation action, this has resulted in a considerable conservation effort being directed towards islands. A large body of empirical data relating to biodiversity conservation in insular situations has resulted.

It is incumbent upon us to sift through the information derived from insular conservation efforts to see what can be applied to the more significant continental and oceanic environments. Such information is likely to be particularly relevant in the former situation as man-induced fragmentation of once-continuous continental ecosystems is increasingly giving rise to quasi-insular situations on the continents.

Oceanic islands are only the most obvious manifestation of the "island" phenomenon: habitat islands occur naturally throughout the biosphere. Mountains located in lowlands, freshwater lakes isolated from one another by terrestrial systems and even fynbos shrublands located in a forest matrix (Bond et al. 1988) all behave like "islands". Even the submerged portions of islands within

[1] WWF South Africa, P O Box 456, Stellenbosch, 7599, South Africa
[2] Percy FitzPatrick Institute of African Ornithology, University of Cape Town, Rondebosch, 7700, South Africa

Ecological Studies, Vol. 115
Vitousek et al. (eds) Islands

Lake Malawi behave like terrestrial islands, each having its own unique assemblage of fish species (Ribbink et al. 1983).

One can even view large systems, such as the Mediterranean-type climate zones on the continents, as being winter-rainfall "islands" within predominantly summer-rainfall continents (Macdonald 1991). Individual river systems are "islands" of lotic freshwater ecosystems, often quite biogeographically distinct from adjacent systems.

It is the "insular", or patchy, nature of the world's environments that has given rise to its biodiversity. Hence it is the most patchy portions of the planet that hold most of this biodiversity, as demonstrated by the contrast between South Africa and Australia. The former is only 1 211 040 km^2 in extent and spans latitudes 22 °S to 34 °S and longitudes 16 °E to 33 °E. The latter is six times larger (7 617 930 km^2) and has much greater latitudinal (10 °S to 38 °S) and longitudinal coverage (113 °E to 153 °E). However, South Africa has much greater topographical relief and, as a result, a much more varied climate. Even though much smaller, South Africa ends up being slightly more diverse (vascular plant species 21 000:17 500, birds 718:656, mammals 227:224). If we are going to retain much of this "island-style" diversity, the conservation lessons from real islands might be highly appropriate.

15.2 Insular Lessons for Conservation

15.2.1 Island Biogeography Theory

The major "lesson" has obviously been that of Island Biogeography Theory (MacArthur and Wilson 1967). Although subject to considerable controversy, the basic extrapolation of the empirically based findings on the effects of island size and island dispersion on species richness to the design of protected area networks (Diamond 1975) has been one of the major inputs of ecological theory to conservation management in recent decades (IUCN 1980). We will not develop this theme further in this chapter, as the subject has been extensively debated and we have no novel insights to add.

15.2.2 Susceptibility of Insular Biotas to Alien Invasions

Although difficult to "prove", it is now fairly well established on the basis of empirical observations that insular biotas are more susceptible to alien invasions than are continental biotas (Loope and Mueller-Dombois 1989). Thus, when the Relative Invasion Index (V) is calculated for 12 islands or island archipelagos for which data are available for the size of their native and invasive alien floras (Table 15.1) the mean value of V is 32.2%. By contrast, the mean value of V for ten continental floras was 12.2%. Although there was overlap in these two data series, the trend is clear; islands have a relatively higher proportion of their floras made up of invasive introduced species.

Table 15.1. Area and Size of the Invasive Alien and Native Vascular Floras in Insular and Continental Situations

Place	Area in km^2	Number of Vascular Plant Species[a] Alien (I)	Native (S)	Relative Invasion Index (V)[c]	Alien Species $/km^2$
Antigua	442	180	900	16.7	0.407
British Isles	241 600	405	1492	21.3	1.68×10^{-3}
Canary Islands	7300	700	1700[b]	29.2	0.096
Channel Islands (California)	900	227	621[b]	26.8	0.252
Faeroe Islands	1400	30	310[b]	8.8	0.021
Galápagos	7900	240	541	30.7	0.030
Guadeloupe	1780	149	1668	8.2	0.084
Hawaii	16 500	880	1100	44.4	0.053
New Zealand	268 670	1570	1790	46.7	5.84×10^{-3}
Reunion	2510	460	675	40.5	0.183
Robben Island	5	38	101[b]	27.3	7.6
Seychelles	260	165	220	42.9	0.635
Tristan Da Cunha Group	160	97	41	70.3	0.606
Continents					
Australia	7 617 930	1750	17 500	9.1	2.30×10^{-4}
Austria	82 730	300	3000	9.1	3.63×10^{-3}
California	411 013	1000	5200	16.1	2.43×10^{-3}
Canada	9 220 970	881	3166	21.8	9.55×10^{-5}
Finland	304 610	120	1250	8.8	3.94×10^{-4}
France	550 100	500	4400	10.2	9.09×10^{-4}
Java[d]	132 170	313	4598	6.4	2.36×10^{-3}
Natal	86 967	314	4506	6.5	3.61×10^{-3}
Poland	304 450	275	2025	12.0	9.03×10^{-4}
South Africa	1 221 040	875	21 000	4.0	7.17×10^{-4}
Spain	499 440	750	4900	13.3	1.50×10^{-3}
Stellenbosch	559	159	617	20.5	0.284
Victoria	227 620	850	2750	23.6	3.73×10^{-3}

[a] All other data extracted from recent publications (Kornás and Medwecka-Kornás 1967; Ross 1972; Boucher 1984; Cooper and Brooke 1986; Macdonald 1991).
[b] = No. of native angiosperm species only (Loope and Mueller-Dombois 1989, Table 10.1).
[c] $V = I * 100/I + S$ (Macdonald et al. 1988a).
[d] Java was considered to be more continental than insular in its characteristics (Macdonald and Frame 1988; Macdonald 1991).

However, this difference in proportion could be more a function of the relatively low number of species native to islands as compared to continents. Unfortunately, most of the continental areas for which data could be found were much larger than the insular areas. Notwithstanding this limitation, the trend appears to be that islands do indeed have higher absolute numbers of invasive alien plant species than do continental areas, even where these latter areas are orders of magnitude larger than the islands. The ratios of number of invasive species per unit area are generally much higher for insular situations (Table 5.1)

The size of an area's alien flora is influenced by many factors other than its area. Thus the location of the area around or adjacent to a large conurbation generally increases the number of alien species present (Kornas 1978; Muir 1983; Rapoport 1991). In the case of nature reserves in southern Africa, distance from urban areas was not shown to have a significant effect on the number of invasive alien plant species, but both mean annual rainfall and the number of visitors to reserves were positively correlated with this statistic (Macdonald et al. 1986). In an analysis of an international data set on nature reserves, the statistical relationship with reserve visitation rates was confirmed as was a general positive relationship with the size of the reserve's native flora (Macdonald et al. 1989). The nature of these relationships varied between biomes (Macdonald et al. 1989).

In order to reduce uncontrolled differences arising from variations in human disturbance, we also provide comparisons of some relatively standardised data on aliens from nature reserves in both insular and continental situations (Fig. 15.1). Although the differences are not clearcut, it appears from this scatter diagram that certain island nature reserves have higher absolute numbers of invasive alien plant species than do continental reserves, some of which are very much larger than the island reserves. Other than this, the general tendency is for island reserves to lie on the upper limit for the observed number of alien plant species in reserves of that size.

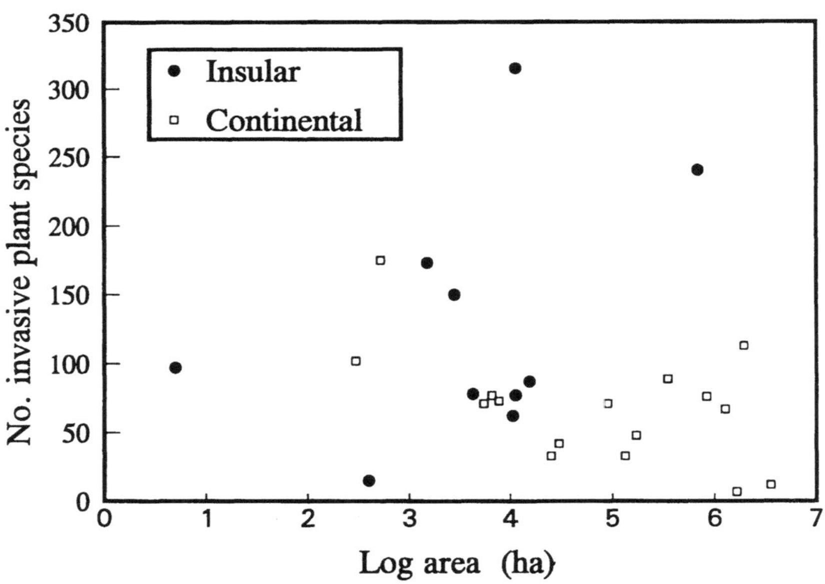

Fig. 15.1. The relationship between the absolute number of invasive alien species and the area of nature reserves on islands and continents. (Data for islands from Devine 1977; Pickard 1984; Brockie et al. 1988; Strahm 1988; data for continents from Ward and Ingwersen 1979; Macdonald et al. 1986, 1988a; Loope et al. 1988; Macdonald and Frame 1988)

The only continental reserves which often have very high absolute numbers of alien species are some small nature reserves located in urban surrounds within Mediterranean-type biomes (Macdonald et al. 1988a). However, the point has been made that these biomes are in fact quasi-insular situations (Macdonald 1991). By contrast, the only islands which regularly do not have large numbers of invasive alien vascular plants are those of high latitudes: for example, the 29 000-ha Marion Island has only ten species (Cooper and Brooke 1986), Macquarie Island Nature Reserve (12 785 ha) only five and Heard Island (37 500 ha) none (Clark and Dingwall 1985). This accords with the observation that as climatic conditions become more extreme, the number of invasive species decreases, for example the entire continent of Antartica has no alien vascular plant species (Macdonald et al. 1989). Just how important climate is in enabling alien plant species to invade, is shown by the 5-ha Mondrain Nature Reserve on Mauritius (Strahm 1988): 195 invasive alien vascular plants are recorded from this small tropical reserve.

The proportion of endangered species which is being threatened as a result of alien invasions provides an indirect measure of the "effect" of these invasions (Macdonald et al. 1989). When this measure was compared for the lists of endangered vertebrates held by the World Conservation Monitoring Centre in 1986, it emerged that islands had significantly higher proportions than continental areas in the case of birds, reptiles and amphibians. It was only because of the abnormally high proportion of Australian mammals threatened by invasions that this did not also apply to this group (Macdonald et al. 1989).

The principle underlying this heightened susceptibility to invasion appears simple: an individual island's biota is based on too small a sub-sample of the global gene pool to have generated robust competitors for every available niche. In many cases this manifests itself in the island biota exhibiting a taxonomic "disharmony" when compared to the biota of climatically similar areas on the continents: often ecologically important taxonomic groups are entirely absent in the island situation (Loope and Mueller-Dombois 1989).

Those few gene combinations which, by chance, arrive on an island are spread too thinly over the total niche space present. In the short term this manifests itself as niche broadening: a species which occupies a relatively circumscribed (or narrow) niche on a continent is observed to occupy a much boarder array of niches when present on an island (e.g. Blondel et al. 1988). In the long term, and given appropriate genetic separation possibilities at the population level, this gene spreading manifests itself as adaptive radiation; a founding species evolves into a number of new species, each occupying a different niche. Classic examples are presented by such diverse taxa as the honeycreepers and fruit flies on the Hawaiian Islands (Carlquist 1970), and Darwin's Finches and the daisy genus *Scalesia* on the Galápagos (Grant 1986; Adsersen 1990). At intermediate time scales and degrees of spatial isolation, niche broadening within a species intergrades into this adaptive radiation, e.g. the *Nesospiza* buntings on Inaccessible Island in the Tristan da Cunha Group (Ryan 1992; Ryan et al. 1994). Just how dynamic this process of adaptation can be has been demonstrated by the studies on changing morphologies in Darwin's Finches in response to annual variations in habitat/food conditions in the Galápagos (Grant and Grant 1990).

However complex and intriguing this phenomenon might be in the field of evolutionary biology, the implications for nature conservation are relatively straightforward: insular species are frequently outcompeted by species that have been honed in the much more exacting biotic communities of the mainlands. That this should be so is readily exemplified by a hypothetical example: if one introduced a suitably adapted mainland woodpecker (of the Dendropicidae, the family that has proven successful in this niche throughout the Old and New World's biogeographical realms) to the Galápagos Archipelago, would one not expect it readily to outcompete the Woodpecker Finch *Camarhynchus pallidus*, which sometimes uses a cactus spine held in its beak to extract its prey from cactus stems? Empirical examples of this type of replacement of insular species by continental species are to be found in the Hawaiian archipelago, where a wide range of introduced bird species has become established in native forests where native species have declined in abundance and, in some cases, have disappeared. However, there have been a number of complicating factors such as introduced diseases, introduced plants and introduced predators involved in the replacement (Scott et al. 1988).

More important in terms of conservation of global biodiversity are the numerous examples where such replacements have occured in quasi-insular situations on the continents. Thus the floating aquatic macrophytes, *Eichhornia crassipes* and *Salvinia molesta*, which evolved in the vast continental-scale freshwater environments of the Amazon Basin, have consistently been able to invade the relatively "insular" freshwater ecosystems of Africa, Asia and Australia (Ashton and Mitchell 1989). These smaller river systems provided inadequate evolutionary opportunities to generate competitors capable of withstanding the invasions of these "super macrophytes".

The invasion of the Mediterranean-climate shrublands of the Cape Province of South Africa, the *Fynbos*, by trees that have evolved in the much larger Mediterranean-climate zones of the Mediterranean Basin and Australia provides another example. The roughly 77 000-km^2 Fynbos Biome can be viewed as being a temperate winter-rainfall "island" located at the southern tip of a predominantly tropical summer-rainfall continent (Macdonald 1991). The amazing radiations of woody plant genera such as *Erica* with 526 species and *Aspalathus* with 245 species within the biome indicates once again, how, in a "quasi-insular" situation, the available genetic resources are stretched thinly over the woody plant niche space. It is thus no surprise that adapted tree species from genera such as *Pinus*, *Acacia* and *Hakea* have been able to invade these ecosystems with remarkable ease. A similar example from a truly insular situation can be seen in the highlands of the Galápagos Islands where the "tree" cover is made up of *Miconia robinsonia*, an endemic species of the family Melastomataceae. Robust tree species from the South American mainland such as the *Psidium guajava* and *Cinchona succiruba* are readily able to supplant *Miconia* (Macdonald et al. 1988b).

One of the corollaries to this superior competitive ability of mainland species is that insular species are almost invariably non-invasive when introduced to mainland situations. Although this principle was not specifically recognised when global overviews of the attributes of invaders were recently conducted for plants (Noble 1989) and vertebrates (Ehrlich 1989), it was implicit in some of the characteristics identified as being associated with successful invaders. Thus, Ehrlich (1989) recognised that a successful invader generally had a "large native range" whereas an unsuccessful invader had a "small native range". Brown (1989), in his global review of vertebrate invasions, did, however, explicitly recognise this principle which he stated as follows: "species that are successful invaders tend to be native to continents and to extensive, non-isolated habitats within continents".

15.2.3 Additions of Species

The addition of species which have no ecological analogues in insular communities often causes large disruptions in community composition and ecosystem function. Examples are provided by the massive disruptions caused by introduced mammalian herbivores on islands which had previously had no such herbivores, by introduced mammalian, reptilian, molluscan and ant predators on previously predator-free islands and by nitrogen-fixing tree species introduced to ecosystems where previously only species lacking this adaptation had been present (Macdonald et al. 1989; Ramakrishnan and Vitousek 1989).

Very often, introduced plant species which have new fuel characteristics relative to the native plants have the ability to change fire regimes. This, in turn, has large effects on species composition and ecosystem function of the invaded ecosystem (Macdonald et al. 1989). However, it is often difficult to say with confidence whether the effects of the invader caused the altered fire regime or were simply a symptom of a man-induced change in the pattern of ignitions.

Another good example of this phenomenon can be deduced from the much greater deleterious impact of introduced rats on temperate oceanic island avifaunas than those of tropical islands (Atkinson 1985). The former held species which did not have an evolutionary conditioning to predation from ground-foraging crabs which are basically restricted to the tropics.

Similar effects can be seen in many quasi-insular situations on the continents. Thus the introduction of the predatory fish *Lates niloticus* to Lake Victoria has caused marked declines in abundance of native fish species, to the extent that whole fisheries have collapsed (Fryer 1991). The socio-economic ramifications of the biological effects of this introduction have included an accelerated deforestation of the lake's shores: the Nile Perch has to be smoked rather than sun-dried as were the fish species that the perch has supplanted in the local fisherman's catch.

The introduction of fire-adapted tree species to the Fynbos Biome has resulted in some of the most pervasive alien invasions of any continental ecosystem in the world (Richardson et al. 1992). The compositional changes these additions of species have brought about include the local extinction of shade-intolerant endemic plants: between 800 and 900 species are estimated to be threatened with extinction as a result of these invasions (Macdonald et al. 1989; Richardson et al. 1992). The ecosystem functions which have been altered by these tree invasions include bio-geochemical cycling, soil hydrology, fire regimes and soil erosion.

The havoc wrought by introduced placental mammalian predators in Australian faunal communities (Macdonald et al. 1989) and the massive landscape-level impacts caused by the herbivory of introduced rabbits (Myers 1986) on this "island continent", also provide classic examples of this phenomenon from continental situations.

15.2.4 The Loss of "Key" Species

The loss of certain "key" species from a community can have a disproportionate effect on community composition. We do not know of any scientifically well-established example of this from an oceanic island, although the most famous "popular magazine" example of the phenomenon is the alleged recruitment failure of the Tambalacoque tree *Calvaria major* on Mauritius following the extinction of its alleged avian disperser, the Dodo *Raphus cucullatus* (Temple 1977). However, the validity of this example has been questioned (Owadally 1979).

There are, however, several examples of the phenomenon from quasi-insular situations on the African mainland. For example, the loss of "biotic resistance" in the form of its entire large-mammal predator community is thought to have been responsible for the Cape Peninsula's susceptibility to invasion by the introduced Himalayan Tahr *Hemitragus jemlahicus* (Macdonald and Richardson 1986). In general, natural areas on the African mainland have proven highly resistant to invasion by alien mammals (Brooke et al. 1986; Macdonald and Frame 1988). Pigs, *Sus scrofa*, for example, have failed to establish successful feral populations in South Africa whereas they have done so successfully over a wide area in Australia (Myers 1986). In the few South African areas where they have managed to become established in small areas, predation of the piglets by leopards *Panthera pardus* has been found to be a major factor limiting their success (Botha 1989).

Another example is provided by rabbits *Oryctolagus cunniculus* which have failed to invade on the South African mainland despite numerous introductions (Smithers 1983) but have become established on several of the southern African offshore islands (Cooper and Brooke 1982) and the Australian mainland (Myers 86). The African mainland is renowned for its high diversity of mammalian and avian predators (Smithers 1983; Macdonald and Gargett 1984). Mammalian predators are absent from the southern African offshore islands. Australian ecosystems have much lower diversities of native predators than comparable ecosystems in Africa (Macdonald and Frame 1988).

Other examples of how the removal of indigenous predation pressure can facilitate invasions are provided by the Cape Fur Seal *Arctocephalus pusillus* and

African Penguin *Spheniscus demersus*, which formerly only bred on offshore islands. However, the large carnivores have been progressively extirpated from the coastal areas of the Cape Province during the historical period (Stuart et al. 1985). As this elimination has progressed, the seal and the penguin have established breeding colonies on the mainland (Macdonald 1992a). In 1989, it was estimated that 69% of total seal pup production occurred at the six mainland colonies established this century (Wickens et al. 1991). The penguin has only started nesting on the mainland since 1980 and only at two localities where human settlements have "insulated" them from the effects of native predators (Crawford et al. 1990). At one mainland penguin colony, a leopard, which came from the nearby mountain region, killed 65 of the 150-strong colony in two nights (Hofmeyr 1987a). The leopard continued to raid the colony intermittently and by the time it was eventually killed 3 months later, only six penguins were present in the colony (Hofmeyr 1987b). It is quite clear that if this species alone had frequent access to the site, the colony would not have been viable.

The converse "experiment" was unwittingly carried out when the offshore Marcus Island in Saldanha Bay was joined to the mainland by a causeway during harbour construction. Mainland mammalian carnivores then gained access to the formerly isolated penguin colony, killing several birds before they were successfully controlled and then excluded by a predator-proof wall (Cooper et al. 1985).

That it was the elimination of "keystone" predators that gave rise to the colonisation of the mainland by the African Penguin, rather than this simply being a consequence of an expanding population being forced to utilise suboptimal sites, is suggested by the observation that the species' overall population was declining at the time these colonisations occurred (Crawford et al. 1990). In the case of the Cape Fur Seal, this does not hold true, as the species was experiencing a rapid population increase at this time. It is not certain whether the ability to utilise mainland breeding sites was not one of the factors involved in bringing about this population increase (Wickens et al. 1991).

As fragmentation of mainland ecosystems occurs as a result of human activities, so, we predict, the importance of this phenomenon will increase. Thus in a recent study of artificially created islands within a man-made impoundment, Dean and Bond (1994) have observed that frugivorous birds have become locally extinct and, as a direct result, the regeneration of mistletoe, *Viscum* spp. on the islands has virtually ceased.

15.2.5 Changing Disturbance Regimes Favour Alien Invasions

The deletion or addition of a predator from a community, as discussed above, can be viewed as a special case of this principle. In this case it is the evolutionary *predation regime* which is changed. However, there are numerous other disturbance regimes which each play a role in shaping the biota of an ecosystem, be it insular or continental.

In many cases, the alteration of insular herbivory regimes by the introduction of mammalian herbivores, (e.g. goats, cattle, sheep, pigs, horses) has given rise to new habitat conditions on the affected islands. Thus grazing-adapted plant invaders have often proliferated in such areas. The replacement of much of the native forest vegetation of the Hawaiian lowlands by pasture grasses to form a derived "savanna" can be viewed as an example of this phenomenon. Similarly, the invasion of the upland bogs of the Hawaiian volcanoes by introduced grasses follows the altered soil-disturbance regime brought about by feral pigs rooting in these soils (Loope et al. 1988a).

Another disturbance regime which has been very severely impacted by man on islands, is that of fire. The more-frequent ignitions that mankind has invariably given rise to on isolated oceanic islands have been instrumental in the invasion of many fire-adapted plant species, e.g. the grass *Andropogon virginicus* on Hawaii (Smith 1985). A similar fire-induced alien plant invasion in a quasi-insular situation is provided by the invasion of Veld Grass *Ehrharta calycina* into the woodlands of Kings Park in Perth, Australia (Macdonald et al. 1988a). However, it is not certain whether the invasion is the result or the cause of the altered fire regime (Macdonald 1989). By contrast, in the quasi-insular situation of Fynbos patches, it was the relative lack of fire in small "islands" that gave rise to the elimination of many fire-dependent native species (Bond et al. 1988).

In all the above cases insularisation leads to an altered disturbance regime. This, in turn, leads to shifts in the competitive relationships of alien and native taxa. It has been postulated that man-induced change in the global environment will result in rapid changes in local "climate regimes" and that these changes will tend, on average, to favour alien organisms over natives, leading to a global enhancement of the phenomenon of alien invasions (Macdonald 1992b).

15.2.6 Alien-Dominated Ecosystems Are Unstable in the Long-Term

Using the argument that it is only alien species that will be able to cope with rapid man-induced changes in the environment, Westman (1990) has proposed that the policy of removing alien organisms from protected areas is ecologically unsound. The final lesson we can deduce from island "experiments" in the functioning of biodiversity, sheds some light on this proposition.

On the island of Reunion in the Indian Ocean a survey of alien vascular plants invading native vegetation remnants was carried out during February 1989, 2 months after the island experienced a major cyclone (Macdonald et al. 1991). Observations indicated that native woody plant species had survived the cyclone with only a slight a seasonal loss of leaves. By contrast, several of the important alien tree invaders of the island, such as *Casuarina equisetifolia* and *Solanum mauritianum* (a South American mainland species, alien to the Mascarene Islands), had shown extensive tree falls and stem breakages. In the case of the widespread invasive *Psidium cattleianum*, large stands had been completely defoliated by the cyclone and were still strikingly bare at the time of our survey.

Mature individuals of the alien tree *Cryptomeria japonica* (which had been extensively planted on the island for timber production) had had their trunks snapped like match sticks over whole hillsides. Another alien timber tree, *Eucalyptus* sp., had been similarly affected. The lesson to be derived from these observations is quite simple: the alien species were all able to grow satisfactorily, and, in some cases, to thrive and outcompete native woody plants *in the short term*. However, when a cyclone, which is an infrequent but highly important selective force characteristic of the environment of the Mascarene Islands, occured, these alien tree species showed themselves to be poorly adapted to the *long-term* environment of these islands.

Another example is provided by alien tree species, such as the Australian *Acacia mearnsii*, which invade riparian vegetation in the Fynbos Biome. It has been observed that invasive alien trees characteristically have lower root/shoot standing crop ratios than do native species (Rutherford et al. 1986), resulting in riparian communities dominated by top-heavy woody plants. When these rivers, which are normally dominated by flood-resistant species such as the giant sedge, *Prionium serratum*, are subjected to flash floods, the invading tree species are ripped out by the floodwaters, often dislodging extensive sedge mats. After the floods, the bare river banks are recolonised by alien species and a cycle is initiated in which river bank erosion rates are accelerated (Macdonald and Richardson 1986).

Another factor which confers decreased long-term stability on these alien-dominated ecosystems is that the alien invaders are often only successful because they are experiencing "ecological release" from their co-adapted predators and pathogens (Macdonald 1992b). An example of the ecological instabilities inherent in such situations is provided by the invasive shrub *Leucaena leucocephala* on the islands of the Pacific. Here the species had been so successful in the short term that it had formed virtually mono-specific stands over whole steep volcanic hillsides, as was observed on the Hawaiian island of Oahu in 1986. In the 1980s, a psyllid bug *Heteropsylla cubana* gained access to many of these islands and its effects on *L. leucocephala* were so severe that whole stands were killed. The ecological implications of this sudden demise of what had become a dominant component of the vegetation on many islands were so serious that an international effort was launched to control this insect (Steppler and Ramachandran Nair 1987; Djogo 1992).

In the Fynbos Biome an Australian member of the Proteaceae, *Hakea sericea*, has proven a highly successful invader of montane areas. If it had not been actively controlled by local conservation authorities, it might have eventually covered virtually every mountainside in the biome (Fugler 1982). Despite this control, it covered 4800 km^2, or 13% of the Mountain Fynbos, 120 years after its introduction. At this stage a fungal pathogen (thought to be a South African species which adapted to *Hakea*) began to attack the alien, giving rise to die-offs which affected large areas (Morris 1982; Richardson and Manders 1985). Fortunately, most of the affected *Hakea* stands were still sufficiently young not to have completely lost their native seed banks. These species are able to regenerate following the death of *Hakea* (Richardson et al. 1992). However, if large stands had been allowed to cover whole mountain sides for, say, a century, and then this disease had wiped out the, by that time, monospecific stands of *H. sericea*, the entire soil mantle of these mountains could have been lost in post-fire erosion.

A final example is provided by *Pinus* invasions of the Fynbos Biome. Numerous pine species are currently invading the Mountain Fynbos and several of these are capable of forming almost mono-specific stands (Richardson et al. 1992). When high intensity fires burn through stands of *Pinus* in these environments, there is a widespread creation of hydrophobic topsoils. This, in turn, gives rise to very high rates of soil erosion in the immediate post-fire period. Such erosion rates, if allowed to persist, would probably result in the total demise of these mountain catchment areas.

15.3 Conclusion

We have attempted to illustrate in this paper how islands can provide salient lessons for the long-term conservation of the Earth's diverse biota. As so much of the conservation biology of island ecosystems is linked to the phenomenon of alien invasion, so too are most of these lessons.

It is our contention that, as the man-induced fragmentation of the formerly continuous continental ecosystems proceeds, the relevance of these island lessons can only increase.

The most important overall lesson is that conservation managers can anticipate an increase in the tempo and intensity of alien invasions of natural areas. It is our main conclusion that such invasions should not be allowed to proceed unchecked, if we are to retain the diversity of native species which will be required for natural areas to meet the challenges of the coming century.

Acknowledgments. Ian Macdonald thanks SCOPE for making his participation in this meeting possible. John Cooper acknowledges financial support received from the South African Department of Environment Affairs.

References

Adsersen H (1990) Intra-archipelago distribution patterns of vascular plants in Galápagos. Monogr Syst Bot Mo Bot Gard 32: 67–78

Ashton PJ, Mitchell DS (1989) Aquatic plants: patterns and modes of invasion, attributes of invading species and assessment of control programmes. In: Drake JA, Mooney HA, di Castri F et al. (eds) Biological invasions: a global perspective. Wiley, New York, pp 111–154

Atkinson IAE (1985) The spread of commensal species of *Rattus* to oceanic islands and their effects on island avifaunas. In: Moors PJ (ed) Conservation of island birds: case studies for the management of threatened island species. Int Council for Bird Preservation. Cambridge, pp 35–81

Blondel J, Chessel D, Frochet B (1988) Bird species impoverishment, niche expansion, and density inflation in Mediterranean island habitats. Ecology 69: 1988–1917
Bond WJ, Midgley J, Vlok J (1988) When is an island not an island?: insular effects and their causes in fynbos shrublands. Oecologia 77: 515–521
Botha SA (1989) Feral pigs in the Western Cape Province: failure of a potentially invasive species. S Afr For J 151: 17–25
Boucher C (1984) An analysis of floristic attributes of Cape Fynbos vegetation at selected sites in South Africa. Poster Paper Presented at Ecological Society of Australia Symp, Sydney
Brockie RE, Loope LL, Usher MB, Hamann O (1988) Biological invasions of island nature reserves. Biol Conserv 44: 9–36
Brooke RK, Lloyd PH, De Villiers AL (1986) Alien and translocated terrestrial vertebrates in South Africa. In: Macdonald IAW, Kruger FJ, Ferrar AA (eds) The ecology and management of biological invasions in southern Africa. Oxford University Press, Cape Town, pp 63–74
Brown JH (1989) Patterns, modes and extents of invasions by vertebrates. In: Drake JA, Mooney HA, di Castri F et al. (eds) Biological invasions: a global perspective. Wiley, New York, pp 85–109
Carlquist S (1970) Hawaii: a natural history. Natural History Press, Garden City, NY
Clark MR, Dingwall PR (1985) Conservation of islands in the Southern Ocean: a review of the protected areas of Insulantarctica. IUCN, Gland, Switzerland
Cooper J, Brooke RK (1982) Past and present distribution of the feral European rabbit *Oryctolagus cuniculus* on southern African offshore islands. S Afr J Wildl Res 12: 71–75
Cooper J, Brooke RK (1986) Alien plants and animals on South African continental and oceanic islands: species richness, ecological impacts and management. In: Macdonald IAW, Kruger FJ, Ferrar AA (eds) The ecology and management of biological invasions in southern Africa. Oxford University Press, Cape Town, pp 133-142
Cooper J, Hockey PAR, Brooke RK (1985) Introduced mammals in South and South West African islands: history, effects on birds and control. In: Bunning LJ (ed) Proc Birds and Man Symp. Witwatersrand Bird Club, Johannesburg, pp 179–203
Crawford RJM, Williams AJ, Randall RM, Randall BM, Berruti A, Ross GJB (1990) Recent population trends of Jackass Penguins *Spheniscus demersus* off southern Africa. Biol Conserv 52: 229–243
Dean WRJ, Bond WJ (1994) Apparent avian extinctions from islands in a man-made lake, South Africa. Ostrich 65: 7–13
Devine WT (1977) A programme to exterminate introduced plants on Raoul Island. Biol Conserv 11: 193–207
Diamond JM (1975) The island dilemma: lessons of modern bio-geographic studies for the design of natural reserves. Biol Conserv 7: 129–146
Djogo APY (1992) The possibilities of using local drought-resistant multipurpose tree species as alternatives to Lamtoro *Leucaena leucocephala* for agroforestry and social forestry in West Timor. Working Paper 32, Environment and Policy Institute. East-West Center, Honolulu pp 1–41
Ehrlich P (1989) Attributes of invaders and the invading process. In: Drake JA, Mooney HA, di Castri F. et al. (eds) Biological invasions: a global perspective. Wiley, New York, pp 315–328
Fryer G (1991) Biological invasions in the tropics: hypotheses versus reality. In: Ramakrishnan PS (ed) Ecology of biological invasion in the tropics. International Scientific Publications, New Delhi, pp 87–101
Fugler SR (1982) Infestations of three Australian *Hakea* species in South Africa and their control. S Afr For J 120: 63–68
Given DR (1992) An overview of the terrestrial biodiversity of Pacific Islands. David Given, Christchurch, New Zealand
Grant PR (1986) Ecology and evolution of Darwin's Finches. Princeton University Press, Princeton
Grant PR, Grant BR (1990) Plant-animals interactions: consumption of seeds by Darwin's Finches. Monogr Syst Bot Mo Bot Gard 32: 179–187
Hofmeyr J (1987a) Bettys Bay penguins and the leopard. Promerops 177: 6–7
Hofmeyr J (1987b) The leopard saga sequel. Promerops 178: 6–7

Honegger RE (1981) List of amphibians and reptiles either known or thought to have become extinct since 1600. Biol Conserv 19: 141–158
IUCN (1980) World conservation strategy. IUCN-UNEP-WWF, Gland, Switzerland
King WB (1985) Island birds: will the future repeat the past? ICBP Tech Publ 3: 3–15
Kornás J (1978) Remarks on the analysis of a synanthropic flora. Acta Bot Slovac Acad Sci Slovac Ser A 3: 385–393
Kornás J, Medwecka-Kornás A (1967) The status of introduced plants in the natural vegetation of Poland IUCN Publ New Ser 9: 38–45
Loope LL, Mueller-Dombois D (1989) Characteristics of invaded islands, with special reference to Hawaii. In: Drake JA, Mooney HA, di Castri F et al. (eds) Biological invasions: a global perspective. Wiley, New York, pp 257–280
Loope LL, Hamann O, Stone CP (1988a) Comparative conservation biologies of oceanic archipelagoes: Hawaii and the Galápagos. BioScience 38: 272–282
Loope LL, Sanchez PG, Tarr PW, Loope WL, Anderson RL (1988b) Biological invasions of arid land nature reserves. Biol Conserv 44: 95–118
MacArthur RH, Wilson EO (1967) The theory of island bio-geography. Princeton University Press, Princeton
Macdonald IAW (1991) Conservation implications of the invasion of southern Africa by alien organisms. PhD Thesis, University of Cape Town, Cape Town
Macdonald IAW (1992a) Vertebrate populations as indicators of environmental change in southern Africa. Trans R Soc S Afr 48: 87–122
Macdonald IAW (1992b) Global change and alien invasions: implications for biodiversity and protected area management. In: Solbrig OT, van Emden HM, van Oordt PGWJ (eds) Biodiversity and global change. IUBS, Paris, pp 197–207
Macdonald IAW, Frame GW (1988) The invasion of introduced species into nature reserves in tropical savannas and dry woodlands. Biol Conserv 44: 67–93
Macdonald IAW, Gargett V (1984) Raptor density and diversity in the Matopos, Zimbabwe. In: Ledger JL (ed) Proc V Pan-Afr Ornithol Congr, Southern African Ornithological Society, Johannesburg, pp 287–308
Macdonald IAW, Richardson DM (1986) Alien species in terrestrial ecosystems of the fynbos biome. In: Macdonald IAW, Kruger FJ, Ferrar AA (eds) The ecology and management of biological invasions in southern Africa. Oxford University Press, Cape Town, pp 77–91
Macdonald IAW, Powrie FJ, Siegfried WR (1986) The differential invasion of southern Africa's biomes and ecosystems by alien plants and animals. In: Macdonald IAW, Kruger FJ, Ferrar AA (eds) The ecology and management of biological invasions in southern Africa. Oxford University Press, Cape Town, pp 209–225
Macdonald IAW, Graber DM, DeBenedetti S, Groves RH, Fuentes ER (1988a) Introduced species in nature reserves in Mediterranean-type climate regions of the world. Biol Conserv 44: 37–66
Macdonald IAW, Ortiz L, Lawesson JE, Nowak JB (1988b) The invasion of the Santa Cruz highlands in Galápagos National Park by the red quinine tree *Cinchona succiruba*: an evaluation of current control methods and some recommendations. Environ Conserv 15: 215–220
Macdonald IAW, Loope LL, Usher MB, Hamann O (1989) Wildlife conservation and the invasion of nature reserves by introduced species: a global perspective. In: Drake JA, Mooney HA, di Castri F et al. (eds) Biological invasions: a global perspective. Wiley, New York, pp 215–255
Macdonald IAW, Thebaud C, Strahm WA, Strasberg D (1991) Effects of alien plant invasions on native vegetation remnants on La Réunion (Mascarene Islands, Indian Ocean). Environ Conserv 18: 51–61
Morris MJ (1982) Gummosis and die-back of *Hakea sericea* in South Africa. In: van de Venter HA, Mason M (eds) Proc 4th Natl Weeds Conf of South Africa. Balkema, Cape Town, pp 51–54
Muir BG (1983) Weeds in national parks of Western Australia. Paper presented at Weeds Conference, Adelaide, 7 pp

Myers K (1986) Introduced vertebrates in Australia, with emphasis on the mammals. In: Groves RH, Burdon JJ (eds) Biological invasions: an Australian perspective. Australian Academy of Science, Canberra, pp 120–136
Noble IR (1989) Attributes of invaders and the invading process: terrestrial and vascular plants. In: Drake JA, Mooney HA, di Castri F et al. (eds) Biological invasions: a global perspective. Wiley, New York, pp 301–313
Owadally AW (1979) The dodo and the Tambalacoque tree. Science 203: 1363–1364
Pickard J (1984) Exotic plants on Lord Howe Island – distribution in space and time, 1853–1981. J Biogeogr 11: 181–208
Ramakrishnan PS, Vitousek PM (1989) Ecosystem-level processes and the consequences of biological invasions. In: Drake JA, Mooney HA, di Castri F et al. (eds) Biological invasions: a global perspective. Wiley, New York, pp 281–300
Rapoport EH (1991) Tropical versus temperate weeds: a glance into the present and future. In: Ramakrishnan PS (ed) Ecology of biological invasions in the tropics. International Scientific Publications, New Delhi, pp 41–51
Ribbink AJ, Marsh BA, Marsh AC, Ribbink AC, Sharp BJ (1983) A preliminary survey of the cichlid fishes of rocky habitats in Lake Malawi. S Afr J Zool 18: 147–310
Richardson DM, Manders PT (1985) Predicting pathogen-induced mortality in *Hakea sericea* (Proteaceae), an aggressive alien plant invader in South Africa. Ann Appl Biol 106: 243–254
Richardson DM, Macdonald IAW, Holmes PM, Cowling RM (1992) Plant and animal invasions. In: Cowling RM (ed) The ecology of Fynbos – nutrients, fire and diversity. Oxford University Press, Cape Town, pp 271–308
Ricklefs RE, Buffetat E, Hallam A et al. (1990) Biotic systems and diversity: report of working group 4, Interlaken workshop for past global changes. Palaeogeogr Palaeoclimatol Palaeoecol 82: 159–168
Ross JH (1972) The flora of Natal. Mem Bot Surv S Afr 39: 1–418
Rutherford MC, Pressinger FM, Musil CF (1986) Standing crops, growth rates and resource use efficiency in alien plant invaded ecosystems. In: Macdonald IAW, Kruger FJ, Ferrar AA (eds) The ecology and management of biological invasions in southern Africa. Oxford University Press, Cape Town, pp 189–199
Ryan PG (1992) The ecology and evolution of *Nesospiza* buntings. PhD Thesis, University of Cape Town, Cape Town
Ryan PG, Moloney CL, Hudon J (1994) Color variation and hybridization among *Nesospiza* buntings on Inaccessible Island, Tristan da Cunha. Auk 111: 314–327
Scott JM, Kepler CB, Van Riper C, Fefer SI (1988) Conservation of Hawaii's vanishing avifauna. BioScience 38: 238–253
Smith CW (1985) Impact of alien plants on Hawaii's native biota. In: Stone CP, Scott JM (eds) Hawaii's terrestrial ecosystems: preservation and management. University of Hawaii, Honolulu, pp 18–250
Smithers RHN (1983) The mammals of the southern African subregion. University of Pretoria, Pretoria
Steppler HA, Ramachandran Nair PK (1987) Agroforestry: a decade of development. International Council for Research in Agroforestry, Nairobi
Strahm W (1988) Mondrain Nature Reserve and its conservation management. Proc R Soc Art Sci Mauritius 5: 139–177
Stuart CT, Macdonald IAW, Mills MGL (1985) History, current status and conservation of large mammalian predators in the Cape Province, Republic of South Africa. Biol Conserv 31: 9–17
Temple SA (1977) Plant-animal mutualism: evolution with Dodo leads to near extinction of plant. Science 197: 885–886
Ward JE, Ingwersen F (1979) A checklist of vascular plant species in the Tidbinbilla Nature Reserve. Conservation Mem No 8, Conservation & Agricultural Dept of the Capital Territory, Canberra
Westman WE (1990) Managing for biodiversity: unresolved science and policy questions. BioScience 40: 26–33
Wickens PA, David JHM, Shelton PA, Field JG (1991) Trends in harvests and pup numbers of the South African Fur Seal: implications for management. S Afr J Mar Sci 11: 307–326

16 Saint Helena: Sustainable Development and Conservation of a Highly Degraded Island Ecosystem

M. MAUNDER[1], T. UPSON[1], B. SPOONER[2], and T. KENDLE[3]

16.1 Introduction: Small Islands and Sustainable Development

Small islands, such as St Helena, provide special challenges for economic development and environmental management. They are often characterised by high levels of environmental degradation and species extinction (Smith et al. 1993) and relatively high levels of social poverty. Their isolation, size, social and settlement history, restricted suite of natural resources, and vulnerability to the external influences of larger economies endow them with special problems (Kristoferson et al. 1985; Towle 1985). Patterns of environmental and economic degradation are of increasing concern in many island systems. For example, both in the Caribbean, and the US Pacific Territories, a post-war decline in agriculture has resulted from large and persistent wage discrepancies encouraging movement of labour from agriculture to tourism (McElroy and de Albuquerque 1990). Underpinning these recent economic changes are long histories of environmental degradation. This is not a recent syndrome; archaeological evidence suggests that some Pacific islands, notably Isla de Pascua, underwent environmental degradation as long ago as 1200 to 800 B.P. (Flenley et al. 1991). In the Caribbean (Watts 1987), Atlantic Islands (Roberts 1989) and Mascarenes (Gade 1985; Cheke 1987), a pattern of massive environmental degradation dates back to European colonial settlement and extensive natural resource exploitation with subsequent development of fragile plantation based economies (Brookfield 1958).

These characteristics were given special regard in the United Nations Conference on Environment and Development (UNCED – The Earth Summit) at Rio de Janeiro in June 1992, when a pledge was made to convene a Global Conference on the Sustainable Development of Small Island Developing Countries. The wider and complex set of principles for sustainable development agreed upon under the *Rio Declaration* are being implemented under the programme known as Agenda 21 (Quarrie 1992). The most significant change in direction of policy and thinking

[1] Conservation Unit, Living Collections Department, Royal Botanic Gardens, Kew, Richmond, Surrey, TW9, 3AB, UK
[2] Independent Consultant, International Institute for Environment and Development, 3 Endsleigh Street, London, WC1H ODD, UK
[3] Department of Horticulture, Plant Science Laboratories, University of Reading, Whiteknights, Reading, RG6 2AS, UK

Ecological Studies, Vol. 115
Vitousek et al. (eds) Islands

arising from Rio are that (derived from The Rio Declaration on Environment and Development):

Human beings are at the centre of concerns for sustainable development.
The right to development must be fulfilled so as to equitably meet developmental and environmental needs of present and future generations.
In order to achieve sustainable development, environmental protection shall constitute an integral part of the development process and cannot be considered in isolation from it.
Environmental issues are best handled with the participation of all concerned citizens.

St Helena offers a case study where an island community has come to the end of a historical series of economic "boom and bust" cycles. The natural resources have been over-utilised, and the future lies in both ecological restoration and a change in social and political mechanisms to promote sustainable development.

The loss and degradation of unique island biotas has been documented, yet the role of the indigenous biota in sustainable development has been given little attention. The characteristically disharmonic and degraded island biotas would appear to offer fewer opportunities for utilisation than a taxonomically more diverse continental biota. What takes priority, the intensive management of degraded fragments of original habitats or the retention or securing of ecological processes to sustain human communities?

16.2 Geology and Climate of St Helena

St Helena lies in the South Atlantic Ocean about 926 km (500 nautical miles) to the east of the Mid-Atlantic ridge (Hoogesteger 1988). The island is the deeply eroded summit of a composite shield volcano (Daly 1927). The island emerged above sea level about 14.3 million years ago; volcanic activity ceased at about 7.5 million years ago (Baker et al. 1967). The island is approximately 122 km^2 in plan view, about 16 km long, 10 km at its widest part.

The island has a remarkably stable sub-tropical climate; this is due to the influence of the South East Trade Wind belt and the Benguela Current. Cloud cover, fog and rainfall have all, at best, been infrequently recorded until recent times, and the data that is available is hard to quantify (Mathieson 1990). The mean annual temperature at Jamestown (25 m a.s.l.) is 24.2 °C, mean annual rainfall ranges from 175 to 1050 mm. There is a widespread conviction amongst older islanders that significant climate change has occurred in their lifetime, Mathieson (1990) could find no conclusive evidence for this idea.

16.3 The Original Ecology of St Helena

The original ecology and vegetation of the island has been almost totally destroyed, today semi-natural habitats occupy less than 1% of the land area. On first discovery, the island was likened to a terrestrial paradise by Portuguese mariners, the high sea cliffs were described as festooned with forest and with numerous freshwater streams. The endemic fauna was dominated by birds and invertebrates. An original avifauna contained two endemic flightless rails, endemic genera of cuckoo, *Nannococcyx psix* and dove, *Dysmoropelia*, and an endemic species of hoopoe, *Upupa antaios* (Westmore 1963; Olson 1973, 1975). Only one endemic bird species, the wire bird (*Charadrius sanctaehelenae*) still survives. It is suspected that the island prior to settlement supported extensive sea bird colonies. There are no native land mammals, reptiles or amphibians. There are no freshwater fish; however, the coastal waters support ten species of endemic fish (Edwards and Glass 1987a,b). The native invertebrate fauna contains at least 200 endemic species of invertebrate (Basilewsky 1970).

The native flora consists of about 60 species of flowering plant and ferns, 50 of which are endemic, with a remarkable 9 endemic genera. Cronk (1989) has reconstructed a hypothetical map of the island's original vegetation. At high altitudes (ca. 700–820 m a.s.l.) the dominant vegetation was influenced by high rainfall and mist interception, and dominated by the endemic tree ferns, *Dicksonia arborescens*, with a rich fern and bryophyte community. Lower on the slopes were cabbage tree (*Lachanodes arborea*) woodlands (600–750 m a.s.l.) a dense low woodland with *Commidendrum spurium*, *Petrobium arboreum*, and *Dicksonia arborescens*. This graded into a drier type of woodland (ca. 500–650 m a.s.l.) dominated by *Commidendrum robustum*, *Trochetiopsis erythroxylon*, *Phylica polifolia*. The main areas of mid-altitude (ca. 300–500 m a.s.l.) were covered by extensive areas of woodland dominated by *Commidendrum robustum* and *Commidendrum rotundifolium*. With decreasing precipitation an open scrub (ca. 100–500 m a.s.l.) community grew dominated by *Trochetiopsis melanoxylon* and *Commidendrum robustum*. On the edge of the island where salt spray and drought prevents tree growth, the vegetation was dominated by the low hummocks of *Commidendrum rugosum* and plants of *Frankenia portulacifolia*. In some areas the salinity of the soil supported a restricted range of halophyte species such as *Suaeda helenae*, *Osteospermum sanctae-helenae*, and *Hypertelis acida*.

16.4 Mechanisms of Environmental Degradation

The historical lack of an integrated approach to development and environmental management is well illustrated by the history of St. Helena. The processes of degradation started before the island was permanently colonised. Goats were introduced by the Portuguese in 1513 (Brooke 1824) to supply meat for victualling ships, by 1588 flocks "near a mile long" were described (Hakluyt 1589). Crews from ships also planted fruit and vegetables and released other livestock. A large number of texts, for instance Coblentz (1978), single out the introduced goat as a major cause of habitat loss on islands. However on St Helena a more complex picture of inter-related factors exists, these include:

Un-managed populations of feral livestock
Clearance of vegetation for crop and pasture development for smallholdings and estates.
Clearance of trees for charcoal and timber for small-scale industry.
Clearance of trees for construction.
Sudden and significant fluctuations of population with associated demands on resources associated with temporary garrisons, whaling and sailing fleets.
Invasive plant species introduced as crops and ornamentals.
Introduced insect pests.
Erosion-prone volcanic soils.
Modified soil processes resulting from forest clearance and possibly the loss of nesting seabird colonies.
Surviving populations of endemics subject to continued threats from inbreeding, stochastic events and invasives/pathogens

There were key periods when high population sizes must have represented a significant drain on the garrison/slave economy and the natural resources of the island. The resident population of the island reached a peak between 1900 and 1902 when the island was used as a prison during the Boer War, when over 6000 prisoners increased the population to nearly 10 000 people.

The limitations of local resources quickly became apparent to the first settlers, and supplies of gumwood became scarce and largely exhausted by the 18th century (Cronk 1986a). The island authorities had to balance the feral and domestic livestock, particularly goats and sheep, as both a resource and environmental liability. In a response to the shortage of timber in the 18th century, a number of conflicts arose over access to timber and forage for feral livestock. In 1744, the island government attempted to impound goats causing damage to valuable ebony trees (*Trochetiopsis melanoxylon*). The Court of Directors of the East India Company in London overturned the island government's decisions, forcing the following statement from the island authorities: "we have repaid the two Messrs Greentree their fines according to your orders, as you are of the opinion that the Goats are of more use here than ebony they shall not be destroyed for the future" (Cronk 1986b).

In common with other oceanic islands, this pattern of over-harvesting of high value plant resources has resulted in dramatic declines in species populations. Examples include the exhaustion of mahogany stocks in the Caribbean, the over-harvesting of endemic palms (Stuessy et al. 1983) and the extinction of the endemic sandalwood *Santalum fernandezianum* on Juan Fernandez (Chile) (Perry 1984). After the over-harvesting of valued resource species the pattern of degradation has been further promoted by plantation monocultures, feral livestock, invasive species and habitat clearance.

Introduced crops have encouraged clearance of key habitat areas. In 1870 Joseph Hooker recommended the planting of *Cinchona* on Diana's Peak (Gosse 1990). At the same time *Phormium tenax* was introduced as a fibre crop (Morris 1884), eventually covering about 1300 hectares. Both crops destroyed one of the best surviving patches of tree fern thicket. The naturalised flora amounts to about 240–270 alien species.

An alien insect, *Orthezia insignis*, the Jacaranda bug, is decimating the last surviving stands of Gumwood, *Commidendrum robustum*. By June 1993 over 100 out of the 2000 trees had been killed by *Orthezia* (Fowler 1993); a biological control programme has been initiated. Introduced pests and pathogens are a problem for island biotas, a graphic example is the near-extinction of the Bermuda cedar (*Juniperus bermudana*) on Bermuda following the accidental introduction of two species of scale insects (Challinor and Wingate 1971). Ironically, this species is now an important forestry species on St Helena and regenerates freely. A number of the introduced insects have a profound economic and social impact; notably the Mediterranean fruit fly (*Ceratitis capitata*) affects domestic fruit production. The termite (*Heterotermes perfidus*) was introduced via the timbers of a broken-up Brazillian slave ship in 1840. The buildings and settlements took a heavy toll, a contemporary describing Jamestown "devastated as by an earthquake".

16.4.1 Extinctions and Habitat Loss

As a direct result of habitat loss and introduced species a number of species have become extinct on the island. All of the endemic birds with the exception of the wire bird are now extinct. Currently, seven vascular plant species are regarded as extinct, with a further two species extinct in the wild; a number of invertebrates have not been seen for a number of decades, most notably the endemic giant earwig *Labidura herculeana*.

A number of surviving endemic plant species have been through dramatic and sometimes prolonged population bottlenecks:

Commidendrum spurium. Two wild plants; an old tree was discovered in 1985 on Cole's Rock, and a seedling in 1985 near Gold Mine Gate.
Commidendrum rotundifolium. Single specimen rediscovered in 1982 on a cliff near Horse Pasture.
Trochetiopsis melanoxylon. Two specimens rediscovered in 1980 on a cliff between Distant Cottage and Asses Ears.
Trochetiopsis erythroxylon. All known stock thought to have descended from one tree at the High Peak that died in the 1960s.
Lachanodes arborea. Thought extinct, circa 60 trees rediscovered at Osborne's Cottage in 1977.
Phylica polifolia. Currently only juvenile 30 plants survive, no surviving mature specimens.
Sium burchellii. Currently about 50 plants are surviving.
Nesiota elliptica. One surviving senescent tree in the wild, four plants held in cultivation on the island.

Plant conservationists face a challenge in managing these dramatically reduced populations. The existing and successful plant propagation run by the St Helenan government needs to expand. This pattern is not unique to St Helena and is becoming an increasingly important issue on a number of island groups, including Puerto Rico (Popenoe 1988), Mascarenes (Owadally et al. 1991), New Zealand (Simpson and De Lange 1993) and Hawaii (David Waugh, pers comm.). A recent workshop on St Helena's endemic flora (Maunder et al. 1993) has reviewed the status of 60 native species. It is recommended that 13 plant species require immediate intensive management, including artificial propagation, to ensure long-term survival.

On the more arid, low productivity areas of the island total devegetation has occurred allowing massive soil erosion and the exposure of raw volcanic material. In places high salinity is a problem, aggravated by exposure to salt laden winds. These degraded lands are collectively known as the Crown Wastes and have expanded to cover 9000 ha of the island's total land area of 12 100 ha. Soil loss in these areas would have been initiated during the early 16th century.

16.5 Species Recovery and Habitat Restoration on St Helena

After a long history of environmental degradation and species loss, the island is entering a period of potential restoration, in terms of both ecology and socio-economic development. Extensive re-introduction programmes have been initiated, and an ambitious soil restoration programme developed. The island has gained a reputation for its remarkable efforts to recover threatened species. Preliminary Recovery Programmes have been drafted (Upson and Maunder 1993) for all native plant species, building upon the success of Agriculture and Forestry Department's Endemic Plants Propagation Programme. In contrast with some Pacific islands (Baines 1991), no traditional environmental management policies exist that can support protected areas or rational land management; the island has been managed since colonisation as a colonial or corporate entity. However the tenacious and enterprising nature of this island community does encourage a pragmatic approach to the land that can easily incorporate the new ethics of sustainable development.

Ecological restoration implies an attempt to reinstate biological communities to a pre-disturbance or colonisation state. This idealistic goal is rarely achieved (Diamond 1987). Far too many components of the original biota are extinct or reduced to remnant and possibly non-viable populations; in addition, large numbers of invasive exotics now dominate the remnant areas of habitat. Although Cronk (1989) has identified the original vegetation distributions, no detailed ecological information on the pre-disturbance condition exists. This is a common problem for island restoration projects; a long history of island degradation often destroys any representative fragments to act as models for restoration (Teytaud 1986; Allen and Wilson 1991).

Any attempt to reinstate a pre-colonisation ecology must be questioned as to whether this is both ecologically possible (Simberloff 1990) and desirable, in a new era of "sustainable development". The conservation of the endemic biota is clearly desirable in terms of retaining the compositional element of an endemic ecosystem; however, will it be the most efficient approach for reinstating a functional ecosystem to support an isolated island community? Considerable expertise in island restoration is developing in New Zealand (Towns and Ballantine 1993), USA (Hansen 1987), Caribbean (Ray 1992) and Mascarenes (Merton et al. 1989). These programmes are led by the need to reinstate threatened species and habitats; they do not include major economic and social components.

Planned restoration will need to move beyond the traditional species focus towards a synthesis where the compositional elements of biodiversity will be supplemented by an urgent need to reinstate ecosystem processes. The scale of the wastelands relative to the island's resources also means that extensive rather than intensive solutions must be found to some of these goals. The need to manage soil erosion and develop a sustainable agricultural landscape will require a focus on the development of ecosystem function rather than pursuing a strict reconstruction of the original ecology. In fact, it could be argued whether the endemic plant species of the island have an important role in a restoration programme dictated by the needs of sustainable development. A direct relationship between species composition and ecosystem function is often assumed by the conservation community; however, this relationship is poorly understood (Walker 1992). For large areas of St Helena, the goals of ecological restoration will focus on ecosystem attributes such as increases in utilisable biomass productivity, soil organic matter, maximum available soil water reserves and length of water availability period.

There has been considerable debate on the value of the endemic flora in intercepting mist (Mathieson 1990, 1992). Currently, mist precipitation recording is based upon an arbitrarily located point capture device that does not indicate the true scale or extent of mist interception. It has been suggested that mist interception may result in a doubling of effective precipitation (Mathieson 1990). A recent report (Anonymous 1990) suggests that increasing the extent of mist-precipitating species in the High Peaks above the 500 m contour to increase mist interception by 10% could result in an increase in stream flow of as great as 30–50%. No evidence exists to indicate that endemic species are any more efficient than introduced species.

Restoration programmes need to be integrated so that their implementation becomes a valued investment for local people and governments struggling in a turbulent economy with scarce resources. The restoration of the eroded lands and the recovery of the island's threatened endemics are only components in the broader picture of ecological restoration and social/economic development for St Helena.

The general focus of the deliberate revegetation policies so far undertaken has been based upon the initial use of drought- and salinity-tolerant pioneer plants, mostly exotic. This is followed, after land stabilisation, with a range of nitrogen-fixing scrub species, especially exotic *Acacia* species, eventually moving to low-grade timber species. The emphasis has been placed upon the most heavily eroding zones, but at the same time natural revegetation is proceeding in many areas following the reduction in feral livestock populations. This new vegetation is often dominated by exotic plant species, some of which have the potential to develop into serious weed problems for the island. Solutions to the management and re-direction of these emergent communities are also needed. Currently, insufficient is known about the dynamics of invasive dominated communities, removal of invasives could both promote or retard ecological restoration. Work is needed to identify those priority species for control (Westman 1990).

To allow long-term commitment to improving the island's environment, the social and economic factors need to be incorporated with environmental issues to develop a national strategy for sustainable development. For example, an important goal would be to develop within the island communities the skills and understanding required for the successful management of dryland forests, something which has never before been demonstrated in the island's history. Restoration therefore cannot be a technical exercise distinct from the wider goals of a Sustainable Environment and Development Strategy. Accordingly, a number of options exist, each of which will be utilised with differing aims (derived in part from Cairns 1991):

1. Retaining original habitats: only feasible in small core areas of remnant habitat, such as tree fern thicket, recognised as priorities for conservation. Labour costs for invasive control will preclude the extensive adoption of this category. Driven by the conservation objectives of reinstating or retaining the original biota and ecosystem function and structure. Priority needs to be given to establishing a linked network of remnant fragments.
2. Restoration designed to protect adjacent ecosystems: restoration of land-linking habitat fragments will focus on preventing invasive species colonisation. Native species will be planted in natural associations. Driven by the conservation objectives of reinstating indigenous biota.
3. Re-establishment of selected economic and ecological attributes: planting of vegetation to provide erosion and water control or timber supply. Driven by economic objectives of establishing vegetation cover and promoting ecosystem processes.
4. Passive recovery: such unmanaged colonisation will be a mixture of native and exotic species. The majority of the islands landscape will be subject to such un-planned recolonisation. Intervention will only be required where a species is a threat to the endemic biota or poses a threat to human welfare or economic activities. The spontaneous colonisation is dominated by exotic species.

16.5.1 Sustainable Environment and Development Strategy for St Helena

Currently, the island of St Helena faces a series of environmental and social challenges. Decline in historical sources of support for the islanders, including in historical times the decline in whaling, the flax fibre industry and the bulb trade. Most recently, the British Nationality Act in 1981 withdrew the right of

St Helenians to live in the UK and in 1991 a new policy was introduced that would phase out the right to obtain work permits for the UK by 1995. Accordingly, an isolated island community with a much degraded set of island resources finds itself increasingly isolated from external sources of support. The island economy is largely supported by UK government subsidy, in the financial year 1989/90 UK budgetary aid totalled $6.5 million, UK Development Aid totalled $3.6 million and UK Technical Co-operation totalled $3.3 million.

National economic development planning mechanisms were established on the island during the 1970s. Major sectoral studies were initiated in the late 1980s and early 1990s. The first draft of a Strategic land Use Plan was submitted in 1993. In spring 1993, the Sustainable Environment and Development Strategy (SEDS) was commissioned by the Overseas Development Administration (ODA) of the UK Government at the request of the St Helena Government. This project is managed by the Conservation Unit, Royal Botanic Gardens, Kew in collaboration with the International Institute of Environment and Development (IIED).

The Terms of Reference for the project required the project team to identify the requirements for institutional strengthening and institutional creation which would allow strategies for economic, social and ecological systems to be combined into a single sustainable development strategy. The mission drew upon the lessons of international experience and the commitments expressed under Agenda 21 of the United Nations Conference on Environment and Development held in Rio de Janeiro in June 1992. In common with many island communities a number of important constraints were identified:

The small size of government departments and the limited numbers of senior staff limits their ability to become involved in the time-consuming activities of developing a sustainable development strategy.
The small population size and the lack of post-school training mean there is no pool of technically trained islanders to support the SEDS programme.
Decision-making at all levels is dominated by the resident UK government authorities.

Following the definition of sustainable development developed by Jacobs and Munroe (1987), the SEDS programme aims to achieve the following:

Making a more effective and sustainable use of natural resources, maintaining ecological function, species diversity, and integrating conservation and development.
Raising awareness of contemporary and potential development issues.
Building a consensus across a broad base of government and public participation about how to set priorities and implement programmes.
Organising partnerships which can build the capacity to act and implement the agreed policies.
Building on existing strengths and developing commitment.
Achieving equity and social justice and social self-determination.

The strategy is a vehicle by which both national incomes and employment can be improved and sustained on a long-term basis because stocks of natural resources can be managed in a sustainable way and because a sufficiently wide range of problems and opportunities can be clearly identified, debated and planned for.

The project has reviewed the following issues: public opinions and concerns, public participation, education and non-government groups, social and economic development, cultural heritage, physical resource management, forestry, marine and coastal strategy, pollution and disasters, biodiversity issues, protected areas, and institutional requirements.

Phase II of the project will follow the recommendations of the island government and include:

Development of Steering Committee and a Working Secretariat on the island, this group to co-ordinate strategy implementation, work on information gathering and policy formulation, stimulate public concern and involvement.
Review legal mechanisms on the island and their impact on the environment.
Maintain and restore populations of endemics to viable levels. Initiate an integrated conservation strategy for the endemic biota, building upon the findings of the recent collaborative workshop between the Captive Breeding Specialist Group of the SSC and RBG Kew (Maunder et al. 1993).
Develop a flexible in situ habitat management approach. Recognising that this will entail managing a dynamic system dominated by invasives.
Initiate a restoration programme for the eroded "Crown Wastes", not as strict restoration programme but as a pragmatic response to limiting resources on the island utilising native and exotic species, for forage and soil restoration.
Develop an integrated pest management programme looking at both agricultural pests and pathogens/pests adversely affecting endemic species.

16.6 Conclusions

St Helena as the ecosystem discovered by mariners in the 16th century is lost and can never be restored; today a new ecology is developing. Under the umbrella of a SEDS programme, it is hoped to develop an island-led balance between species conservation and resource development that sustains the inhabitants of the island. The future of St Helena as an inhabited island and viable community, (rather than as a biological reserve managed primarily for the conservation of an endemic biota), lies in the recognition of a series of mutually interdependent subsystems – economic, sociodemographic, political and ecological. The interactions of these subsystems influencing society as a result of external and internal changes (McElroy and de Albuquerque 1990).

Recognising the integrated nature of ecological and socio-economic systems, the island's ecological fragility, and the need for a diversified economy, several new objectives will need to be introduced. A major effort must be made to restore degraded natural resources, agricultural policy must be reviewed. The latter should include policy that encourages agroforestry. This system is well fitted to the needs of St Helena, through encouraging nutrient recycling and reducing erosion. Historically, the influences of external economic forces have had major impacts on St Helena's economy; accordingly, defences against external free market forces may have to be considered. These could include restrictive land and marine use policies such as the zoning of agricultural and fisheries activities. The success of these activities is dependent on developing a participatory and democratic process for decision making.

St Helena is an example of an island that has passed through episodes of degradation; it is now entering a period of potential restoration. It is a scenario of international relevance; other islands share its fate. The environmental degradation on the Cape Verde Islands has been so severe that evacuation of the resident population has been discussed (Timberlake 1985). The rebuilding of St Helena may prove pertinent as environmental degradation and social pressures will force development-directed ecological restoration in densely populated and degraded islands such as the Philippines or the drier islands of Indonesia. Original ecologies may be irreparably lost, the paradigm of "recombinant ecology" (Soulé 1990) may offer a path to sustainable development through reinstating ecosystem function, but not necessarily the original associations of biodiversity.

Acknowledgments. The authors wish to thank the people of St Helena and Ascension for their patience, good nature and support. His Excellency Governor of St Helena, A. Hoole, and the Chief Secretary, J. Perrot, have shown a keen interest in the project and greatly facilitated its progress. The staff of the Agriculture and Forestry Department have contributed greatly to the project. Tribute is given to George Benjamin for his untiring work in conserving the island's endemic flora. The project was undertaken on behalf of the St Helena Government under assignment from the Overseas Development Administration.

References

Allen RB, Wilson JB (1991) A method for determining indigenous vegetation from simple environmental factors, and its use for vegetation restoration. Biol Conserv 56: 265–280

Anonymous (1990) St Helena water plan, 1990–2010. Public Works and Services Department. St Helena Government, St Helena

Baines GBK (1991) Asserting traditional rights: community conservation in the Solomon Islands. Cult Surviv Q 15(2): 49–52

Baker I, Gale NH, Simons J (1967) The geochronology of the Saint Helena Volcanoes. Nature 215: 1451–1456

Basilewsky P (1970) La Faune Terrestre de L'Ile de Sainte-Helene (Premiere partie). Ann Mus R Afr Cent 181

Brooke TH (1824) A history of the Island of St Helena. 2nd edn. Kingsbury, Padbury and Allen, London, UK

Brookfield HC (1985) Problems of monoculture and diversification in a sugar island: Mauritius. Econ Geogr 34: 25–40

Cairns J Jr (1991) The status of the theoretical and applied science of restoration ecology. Environmental Professional 13: 186–194

Challinor D, Wingate DB (1971) The struggle for survival of the Bermuda cedar. Biol Conserv 3: 220–222

Cheke AS (1987) An ecological history of the Mascarene islands, with particular reference to extinctions and introductions of land vertebrates. In: Diamond AW (ed) Studies of Mascarene island birds. Cambridge University Press, Cambridge, UK
Coblentz BE (1978) Effects of feral goats (*Capra hircus*) on island ecosystems. Biol Conserv 13: 279–286
Cronk Q (1986a) The decline of the St Helena Gumwood, *Commidendrum robustum*. Biol Conserv 35: 173–186
Cronk Q (1986b) The decline of the St Helena ebony, *Trochetiopsis melanoxylon*. Biol Conserv 35: 159–172
Cronk Q (1989) The past and present vegetation of St Helena. J Biogeogr 16: 47–64
Daly RA (1927) The geology of Saint Helena Island. Proc Am Acad Arts Sci 62(2): 31–92
Diamond J (1987) Reflections on goals and the relationship between theory and practice. In: Jordan WR, Gilpin ME, Aber JD (eds) Restoration ecology. Cambridge University Press, Cambridge
Edwards AJ, Glass CW (1987a) The fishes of St Helena Island, South Atlantic Ocean. I. The shore fishes. J Nat Hist 21: 617–686
Edwards AJ, Glass CW (1987b) The fishes of St Helena Island, South Atlantic Ocean. II. The pelagic fishes. J Nat Hist 21: 1367–1394
Flenley JR, King ASM, Jackson J, Chew C, Teller J, Prentice ME (1991) The late Quaternary vegetation and climatic history of Easter Island. J Quat Sci 6: 85–115
Fowler SV (1993) Report on a visit to St Helena. International Institute of Biological Control, Silwood Park, UK (unpubl)
Gade DW (1985) Man and nature on Rodrigues, tragedy of an island commons. Environ Conserv 12(3): 207–216
Gosse P (1990) St Helena 1502–1938. Anthony Nelson Ltd, Oswestry, Shropshire, 447 pp
Hakluyt R (1589) The Principal Navigations, Voyages and Discoveries of the English Nation. 1st edn facsimile with an introduction by DB Quinn and RA Skelton, 1965. Cambridge University Press, Cambridge, UK
Hansen B (1987) Santa Cruz: an island reborn. Nat Cons News June/July: 9–14
Hoogesteger J (1988) The potential for offshore fisheries in the St Helena Exclusive fishing zone. Final report on an ODA Resource Assessment Survey, 1985–1988. Overseas Development Administration, London, UK
Jacobs P, Munroe D (eds) (1987) Conservation with equity: strategies for sustainable development. Proc Conf on Conservation and Development: Implementing the World Conservation Strategy, Ottawa, Canada, IUCN, Gland, Switzerland
Kristoferson L, O'Keefe P, Soussan J (1985) Energy in small island economies. Ambio 14(4–5): 242–244
Mathieson IK (1990) The agricultural climate of St Helena. WS Atkins, Cambridge, UK
Mathieson IK (1992) Irrigation advisors report to the St Helenan Government. WS Atkins, Cambridge, UK
Maunder M, Pearce-Kelly P, Mace G, Clarke D, Upson T, Seal US (1993) Conservation assessment and management plan for St Helena. Captive Breeding Specialist Group/of IUCN, Apple Valley, Minnesota
McElory J, de Albuquerque K (1990) Managing small island sustainability: towards a system design. In: Beller W, D'Ayalu P, Hein P et al. (eds) Sustainable development and environmental management in small islands. Parthenon, New York, USA
Merton D, Atkinson IAE, Strahm W, Jones, C, Empson RA, Mungroo Y, Dulloo E, Lewis R (1989) A management plan for the restoration of Round Island, Mauritius. Jersey Wildlife Preservation Trust and the Ministry of Agriculture, Fisheries and Natural Resources, Mauritius
Morris D (1884) Report upon the present conditions and prospects of the agricultural resources of the island of St Helena (London). Colonial Office, London, UK
Olson SL (1973) Evolution of the rails of the South Atlandic Islands (Aves: Rallidae). Smithson Contrib Zool 152: 1–53
Olson SL (1975) Paleo-ornithology of St Helena Island, South Atlantic Ocean. Smithson Contrib Paleobiol 23: 1–49

Owadally AW, Dulloo ME, Strahm W (1991) Measures that are needed to help conserve the flora of Mauritius and Rodrigues in ex situ collections. In: Tropical botanic gardens: their role in conservation and development. Academic Press, London, UK
Perry R (1984) Juan Fernandez Islands: a unique botanical heritage. Environ Conserv 11(1): 72–76
Popenoe J (1988) One of the world's rarest species. Plant Conservation, a publication of the Center for Plant Conservation 3(4): 6
Quarrie J (ed) (1992) Earth Summit 1992. The Regency Press, London, UK
Ray G (1992) Point of contact: the West Indies. Restor Manag Notes 10(1): 4–8
Roberts N (1989) The Holocene: an environmental history. Blackwell, Oxford, UK
Simberloff D (1990) Reconstructing the ambiguous: can island ecosystems be restored? In: Towns DR, Daugherty CH, Atkinson IAE (eds) Ecological restoration of New Zealand islands. Conservation Sciences Publ No 2. Department of Conservation, Wellington, New Zealand
Simpson P, De Lange PJ (1993) Saving plants by growing them in gardens. In: Froggat P, Oates M (eds) People, plants and conservation. Royal New Zealand Institute of Horticulture, Wellington, New Zealand, pp 99–105
Smith FDM, May RM, Pellew R, Johnson TH, Walter KR (1993) How much do we know about the current extinction rate? TREE 8(10): 375–378
Soulé M (1990) The onslaught of alien spècies and other challenges in the coming decades. Conserv Biol 4(3): 233–239
Stuessy TF, Sanders RW, Matthei OR (1983) *Juania australis* revisited in the Juan Fernandez Islands, Chile. Principles 27(2): 71–74
Teytaud R (1986) Environmental factors and potential natural vegetation in St John, US Virgin Islands. Report to the National Parks Service, Washington, DC, USA
Timberlake L (1985) Africa in Crisis: the causes, the cures of environmental bankruptcy. Earthscan Publications (IIED) London, UK
Towle E (1985) The island microcosm. In: Clark J (ed) Coastal resource management: development case studies. National Park Service Washington, DC, USA
Towns DR, Ballantine WJ (1993) Conservation and restoration of New Zealand island ecosystems. TREE 8(12): 452–457
Upson T, Maunder M (1993) Status of the endemic flora and preliminary recovery programmes. Overseas Development Administration and Royal Botanic Gardens, Kew (unpubl), UK
Walker BH (1992) Biodiversity and ecological redundancy. Conserv Biol 6: 8–23
Watts D (1987) The West Indies: patterns of development, culture and environmental change since 1492. Cambridge University Press, Cambridge, UK
Westman WE (1990) Park management of exotic plant species: problems and issues. Conserv Biol 4(3): 251–254
Westmore A (1963) An extinct rail from the island of St Helena. Ibis 103b: 379–381

Section E
Where Can We Go From Here?

17 Biodiversity and Ecosystem Function: Using Natural Attributes of Islands

R. D. BOWDEN

17.1 Introduction

Islands and island systems offer a number of gradients, attributes, and natural experiments that facilitate examination of relationships between biodiversity and ecosystem function. First, they are often simpler systems than continental areas, reducing the number of complicating variables. Numerous studies, for example, have shown that taxonomic diversity declines strongly with decreasing island size and increasing distance from continental areas (Ash 1992; MacArthur and Wilson 1967; Woodroffe 1987). Even though genetic diversity within species on islands is often high, the overall taxonomic diversity is usually low (Mueller-Dombois 1990). Second, a number of archipelagos contain islands that have similar geologic histories or features (e.g., the Bahamian islands) that reduce complexities in soil age and processes. This can greatly simplify studies of ecosystem processes where highly variable soil conditions make it difficult and sometimes impossible to understand ecosystem functions. Third, and importantly, islands offer a wide gradient of diversity, with natural variation in factors such as latitude, size, age, spatial arrangement, and pattern of disturbance. Many of these factors, notably size and distance from continents, strongly influence biodiversity. Use of these natural gradients and the resulting gradients in biodiversity offers the prospect of establishing natural experiments to examine relationships between biodiversity and functioning within ecosystems. This chapter will offer several examples of relationships between biodiversity and ecosystem function that might be examined by using islands and island systems.

17.2 Using Island Gradients

Natural gradients in taxonomic diversity in island systems provide unique opportunities to examine linkages between biodiversity and ecosystem function. In the south Pacific, for example, mangrove diversity declines dramatically with distance from the southeastern portion of the Asian continent. In New Guinea,

Allegheny College, Department of Environmental Science, Meadville, Pennsylvania 16335, USA

Ecological Studies, Vol. 115
Vitousek et al. (eds) Islands

there are 30 species of mangroves; in Vanuata and New Caledonia there are 10, in Fiji 7, and in Samoa there are 3. In Hawaii there are no native mangrove species (Woodroffe 1987). Using information of this kind, it would be possible to select mangrove ecosystems of varying diversity, and to then examine selected ecosystem properties. Is there a decline in net primary production (NPP) with the decline in mangrove diversity? Is there a relationship among mangrove diversity, litter decomposition, and rates of nutrient cycling? Does mangrove diversity influence avian diversity and seed dispersal, or avian diversity and energy flow through higher trophic level organisms (e.g., avian or reptilian predators)? One might also look at landscape-level effects on diversity, examining the influence of mangrove diversity on adjacent coastal aquatic communities. Do the richer mangrove ecosystems support greater varieties of tropical fish and invertebrates? Does this result in greater energy capture and nutrient flow?

These gradients in taxonomic richness might facilitate additional explorations. Is there a relationship, for example, between tree species richness and rates of organic matter accumulation, mineral weathering, or pedogenesis? Is there a decline in soil invertebrate or fungal diversity that results in declines in the rate of organic matter processing or rates of nutrient retention? Are there differences among resident avian populations that govern net nutrient imports or exports? Are there differences in consumer populations (such as *Anolis* lizards in the Bahamas) that impact trophic structure?

Connections between biodiversity and function might also be tested by employing the already established relationship between island size and biodiversity. A carefully selected archipelago containing islands of different sizes but similar ages, substrate, and history would be expected to yield a set of islands differing in biodiversity, and many of the questions just described could be examined along a gradient of island size rather than island distance. Size, however, might offer additional opportunities for study that are not possible using the gradient of island distance. For example, how does the rate of nutrient input differ among differently sized islands? On Fire Island, a barrier island off the coast of New York (USA), Art et al. (1974) found that oceanic sources were extremely important vectors of nutrient inputs. Selection of differently sized islands might provide a set of ecosystems with different rates of nutrient inputs, enhancing investigations of relationships among nutrient inputs, floral and faunal diversity, and energy and nutrient flows.

Controlled gradients on individual islands might also prove advantageous. For example, climate change could be examined through selection of sites along elevation gradients on active volcanic islands (e.g., Hawaii). Here, excellent records of individual lava flows provide a sequence of sites that have the same geological substrate, but changes in elevation provide a gradient in climate, and hence an ability to examine the role of climate in controlling ecosystem structure and function. The utility of this gradient in examining ecosystem processes has already been demonstrated in a recent study on organic matter and nutrient accumulation (Vitousek et al. 1992).

17.3 Diversity, Disturbance, and Stability

There is no doubt that global environmental change is subjecting ecosystems worldwide to new or increasing disturbances. What is less clear, however, is how ecosystems will respond to those disturbances. One concept that could be explored using island systems is the long-standing controversy concerning ecosystem diversity and stability. There is still considerable disagreement concerning the role of biodiversity in an ecosystem's ability to resist disturbance, as well as the importance of diversity in an ecosystem's resilience following disturbance (May 1974; McNaughton 1988, 1993; Tilman and Downing 1994; van Dobben and Lowe-McConnell 1975). Defining these relationships would be valuable in assessing interactions between diversity and global environmental change. However, establishing experiments to test these relationships has proven difficult due to the inability to control, or sometimes even to define, numerous variables. Islands may be a useful tool for testing these hypotheses concerning diversity, resistance, and resilience. Archipelagos containing islands similar in geological history and exposure to a common disturbance could be selected to provide gradients of biodiversity, size, and spatial arrangement. For example, the Bahamian islands are remarkably similar in geologic history and substrate; the islands vary greatly in area, thus diversity among these islands is likewise expected to vary. These islands are also subjected to a common, large-scale disturbance – hurricanes – that track regularly through the Carribbean.

Examination of island biodiversity and ecosystem function before and after passage of a major hurricane would provide a natural experiment on the role of biodiversity on both ecosystem resistance and resilience following intense disturbance. Variables that could be examined include, but are not limited to, rates of vegetation regrowth and NPP, nutrient inputs, cycling rates, and retention, the return of resident fauna, and energy flow through the trophic structure. Other island systems, such as those in the South Pacific (e.g., Solomon Islands, Philippines) that are exposed to intense tropical cyclones, may also prove suitable for these explorations.

Island systems exhibit great potential in addressing the issue of ecosystem fragmentation, a topic of great concern in the field of conservation biology, where questions of minimum habitat size and habitat diversity are often asked, but where answers often prove difficult and elusive. In recent decades, it has become clear that disturbance and dynamic change are integral to ecosystem development, structure, and function. Landscapes are rarely homogeneous. Rather, they are a dynamic mosaic of patches (Bormann and Likens 1979), such as tree-fall gaps in forests, lightning-induced firescars in grasslands, or wave-damaged sections of coral reefs, all of different ages and stages of development. This diversity of patches within the landscape will also impact the diversity of biotic populations within that landscape. A critical point in the landscape fragmentation issue is that there may be some threshold fragment size at which the size and diversity of metapopulations (populations across the larger landscape level) of flora or fauna remain relatively constant, even though subpopulations within that metapopulation may naturally display drastic fluctuations, in both numbers and diversity, within smaller spatial scales. Threshold fragment sizes for metapopulations may be small for some species, such as soil bacteria, but large for trees in species-rich forests, such as the cyclone-prone south Pacific.

It is not clear how metapopulations will respond to increased levels of disturbance in a highly fragmented landscape, nor what the ultimate impact will be upon ecosystem function. For example, if several subpopulations are decimated, are there sufficient numbers and diversity of remaining individuals or subpopulations to preserve the size and genetic diversity of the metapopulation? What will be the impacts upon ecosystem functions if there is a decline in the metapopulation? An example of this issue is illustrated by the role of timber wolves and moose on the island of Isle Royale in Lake Superior. Following an extremely high wolf density in the late 1970s (Peterson and Page 1988), the population on this 550 km^2 island has crashed to approximately 10% of its peak density (Wayne et al. 1991), and if the remaining subpopulation does not recover, impacts upon vegetation and biogeochemical cycles by moose, the wolves' preferred prey, are likely to be great (Pastor et al. 1988). If the remaining subpopulation does not recover, it may be some time before the metapopulation is able to recolonize the island.

Testing relationships among fragment size, diversity, disturbance, and ecosystem recovery could be accomplished using carefully selected archipelagos containing similarly sized islands that vary in spatial relationships among one another. For example, islands located in close proximity (i.e., short corridors) might show a much higher degree of species overlap or interactions among subpopulations, and hence a more rapid recovery from disturbance, than similarly sized islands spaced over a much larger area (i.e., long corridors). Archipelagos that may be useful for these studies include those in the Caribbean and in the South Pacific, where hurricanes and cyclones are common. Following passage of these tropical storms, do the more closely spaced and more diverse systems recover net primary production more quickly than widely spaced and less diverse systems? Do populations on widely spaced islands recover slowly because of reduced exchange rates with subpopulations on neighboring islands? If a particular species on an island is slow to recover, are there other species that fill in the gap, or allow some degree of redundancy, particularly on the more diverse islands?

17.4 Continental Islands: Terrestrial and Aquatic

Using islands to explore biodiversity-ecosystem function relationships could also be evaluated by expanding our studies from oceanic island systems to a broader definition of island systems. On active volcanic islands, for example, forest fragments (kipukas) surrounded by lava flows vary considerably in size, age, and spatial arrangement. These kipukas can be considered as islands upon islands, and have already been used effectively to examine patterns of genetic diversity

(Carson and Kaneshiro 1976). They may also prove invaluable in studies of population fragmentation and recolonization following disturbance (Carson and Templeton 1984), and could be used to explore many of the concepts mentioned in this chapter. For example, does the distance of these kipukas from the contiguous forest alter avian communities.? What effect, if any, does this have upon seed predation, dispersal, and vegetation composition? Are there differences in floristic composition that impact ecosystem-level processes? Does kipuka size or distances among kipukas impact ecosystem processes? Further examples of "mainland islands" include desert streams, which often contain disjunct fish populations (Turner 1974; Soltz and Naiman 1978), Florida Everglades hammocks (stands of trees raised slightly above the surrounding grass and flowing water system), and high altitude ecosystems (Carson and Templeton 1984).

17.5 Summary

Even though islands comprise a relatively small portion of the earth's land surface area (0.6%, excluding Australia), they offer a number of useful opportunities for exploring relationships between biodiversity and ecosystem function. Natural gradients in biodiversity, geology and geography, and spatial relationships provide valuable natural experiments that allow effective exploration of the influence of biodiversity on ecosystem function. It would be inappropriate to consider islands simply as small-scale versions of continents, but islands are likely to be useful tools where analogous studies are difficult or impossible to establish on mainland areas, and where linkages are subtle and hence difficult to observe.

References

Art HW, Bormann FH, Voigt GK, Woodwell GM (1974) Barrier island forest ecosystem: role of meteorologic inputs. Science 184: 60–62

Ash J (1992) Vegetation ecology of Fiji: past, present, and future perspectives. Pac Sci 46:111–127

Bormann FH, Likens GE (1979) Pattern and process in a forested ecosystem. Springer, Berlin Heidelberg New York

Carson HL, Kaneshiro KY (1976) *Drosophila* of Hawaii: systematics and ecological genetics. Annu Rev Ecol Syst 7:311–345

Carson HL, Templeton AR (1984) Genetic revolutions in relation to speciation phenomena: the founding of new populations. Annu Rev Ecol Syst 15: 97–131

MacArthur RH, Wilson EO (1967) The theory of island biogeography. Princeton University Press, Princeton, NJ

May RH (1974) Stability and complexity in model ecosystems. Princeton University Press, Princeton, NJ

McNaughton SJ (1988) Diversity and stability. Nature 333: 204–205

McNaughton SJ (1993) Biodiversity and function of grazing ecosystems. In: Schulze ED, Mooney HA (eds) Biodiversity and ecosystem function. Ecological Studies, Vol. 99. Springer Berlin Heidelberg New York, pp 361–383

Mueller-Dombois D (1990) Impoverishment in Pacific island forests. In: Woodwell GM (ed) The Earth in transition: patterns and processes of biotic impoverishment. Cambridge University Press, Cambridge, NY pp 199–210
Pastor J, Naiman RJ, Dewey B, McInnes PF (1988) Moose, microbes and the Boreal Forest. Bioscience 38: 770–772
Peterson RO, Page RE (1988) The rise and fall of Isle Royale wolves, 1975–1986. J Mammal 69: 89–99
Soltz DL, Naiman RJ (1978) The natural history of fishes in the Death Valley system. Nat Hist Mus Los Ang Cty Sci Ser 30: 1–76
Tilman D, Downing JA (1994) Biodiversity and stability in grasslands. Nature 367: 363–365
Turner BJ (1974) Genetic divergence of Death Valley pupfish species: biochemical versus morphological evidence. Evolution 28: 281–294
van Dobben WH, Lowe-McConnell RH (eds) (1975) Unifying concepts in ecology. Dr W Junk, The Hague
Vitousek PM, Aplet G, Turner D, Lockwood JJ (1992) The Mauna Loa environmental matrix: foliar and soil nutrients. Oecologia 89: 372–382
Wayne RK, Lehman N, Girman D, Gogan PJP, Gilbert DA, Hansen K, Peterson RO, Seal US, Eisenhawer A, Mech LD, Krumenaker RJ (1991) Conservation genetics of the endangered Isle Royale gray wolf. Conserv Biol 5: 41–51
Woodroffe CD (1987) Pacific Island mangroves: distribution and environmental setting. Pac Sci 41:166–185

18 Experimental Studies on Islands

J.J. EWEL[1] and P. HÖGBERG[2]

18.1 Introduction

As we seek to understand the relationship between biological diversity and ecosystem processes, we often turn to islands because of their unique biota – impoverished, disharmonic, and alien-infested though it may be, there are still those who love it. Islands can be viewed as natural experiments, and ecosystem processes thereon can be compared with those on continents or other islands having similar geology, topography, and climate. Such comparisons should do much to elucidate the relationships between biota and fluxes of materials and energy, as described in Chapter 14.

Nevertheless, island-continent comparisons and natural experiments (Bowden, Chap. 17, this Vol.) are not the only motivations for turning to islands to evaluate biological diversity and ecosystem processes. For a number of reasons, islands lend themselves to experimentation, and the description of such experiments was the task of this working group.

18.2 Advantages of Islands

First among the attractions of performing experiments on islands is the fact that island biota are often well known, and this makes it possible to evaluate gains and losses of species using censuses. Such censuses have been used many times during the development of island biogeography as well as the polemics associated with their applicability to conservation biology.

A second attraction of islands as locales for experiments concerns the opportunity they afford for assessing the functional consequences of harmony and disharmony in their biological spectra (i.e., the degree to which the frequency distribution of taxa on an island corresponds to that of its donor continents). The characteristically skewed spectra of island biota can arise in two ways. In some cases, the biota is disharmonic because some representatives never arrived. In other cases, biological spectra have been modified, often by extinctions associated with the arrival of humans (see James, Chap. 8, this Vol.). In any event, island communities could be subjected to additions and removals at any of several

[1] Department of Botany, Bartram Hall, University of Florida, Gainesville, Florida 32611, USA (*Present Address:* Institute of Pacific Islands Forestry, 1151 Punchbowl St., Rm. 323, Honolulu, Hawaii 96813, USA)

[2] Department of Forest Ecology, Swedish University of Agricultural Sciences, S 90183 Umea, Sweden

Ecological Studies, Vol. 115
Vitousek et al. (eds) Islands

levels – species, life-forms, functional groups, or trophic levels. Communities might be reassembled using species from a variety of sources. One such source could be the ecotypes or functional groups that have developed within taxa that have undergone ecological release, such as the multiple forms of a single species (e.g., *Metrosideros polymorpha*) or a larger taxon, such as a family (e.g., Asteraceae; see Eliassòn, Chap. 4, this Vol.). Depending on the objectives of the experiment, reassembly could be restricted to certain groups, such as indigeneous species or utilitarian species. Still another approach would be to run a disharmony experiment in reverse; this would involve leaving the island biota as-is, and experimentally creating, through selective removal of species, equivalent disharmony on a continental system. Responses would be evaluated both by measuring the consequences of the treatment on the continent and by making island-continent comparisons.

A third attraction of island experiments is that they lend themselves to measurement of ecosystem processes. Fluxes of species, mentioned above, are one category of response that can be measured relatively more easily on islands than on continents. The same is true of biogeochemical linkages between contiguous ecosystems. As one example, fluxes of materials across the land-sea interface have important impacts on islands because of their high ratio of edge to surface area. For another, many low-elevation islands are underlain by limestone that contains freshwater lenses, and the reciprocal flows of water, nutrients, and contaminants between the surface and these pockets of freshwater are more readily measured here than on more geologically complex continents.

Finally, despite their common characterization as abused landscapes, the fact remains that some islands – particularly those that are very remote, very small, and free of freshwater – are as pristine as any equal-sized territory on a continent. Clearly, experiments to be performed on these microcosms should be designed to ensure that they do not have lasting impacts on the natural state that makes them such attractive targets for research in the first place.

18.3 Examples

Nine categories of experiments are briefly outlined in Table 18.1 and in the paragraphs below. These sample experiments (not defined here with proposal-quality rigor) are included as guideposts to the kinds of issues that would lend themselves to experimental science on islands.

18.3.1 Within-Ecosystem Processes

Some experiments on islands would lead primarily to insights about the relationship between biological diversity and ecosystem processes in the ecosystem subjected to manipulation. As an example, under the general topic of making amends for extinctions, scientists might choose substitutes for extinct species based on a number of criteria, some of which include geographic proximity,

Table 18.1. Examples of the kinds of issues related to biological diversity and ecosystem processes that might be addressed by experiments on islands

General topic	Prediction	Experiment	Biodiversity components	Ecosystem processes
Making amends for extinctions	The ecological role of an extinct species can be reinstated through introduction of a functional, rather than a taxonomic, replacement	Introduction of replacement species on paired islands – one with complete range of biota and functions, one with both degraded	Multispecies interactions such as pollinations, seed distribution, predation, and symbioses	Primary productivity, vegetation dynamics, nutrient and hydrologic cycles, and organic matter dynamics
Loss of functional groups	Productivity will decline with removal of plants that are either habitat specialists or generalists, with specialists responding most at gradient extremes and generalists most midway along the gradient	At several sites along an environmental gradient do two treatments involving removal of (1) dominant habitat generalists, and (2) dominant specialists (plus controls for removal disturbance and no removal)	Functional groups related to environment	Productivity at levels of community and functional groups
Volcanic soil age gradients	Impacts of introduced nitrogen-fixing species are inversely proportional to substrate age and proportional to archipelago distance from continental landmasses	Plant or eliminate exotic N_2-fixer on soils of different age, e.g., different lava flows (same island or archipelago), and on archipelagos varying in isolation from continental landmasses	Nodule-inducing microbes; herbivores; native flora	Nitrogen cycling; growth and survival of coexistent plant species
Islands as nutrient concentrators	Manuring by seabird droppings increases primary productivity	On island edges, select pairs of plots that vary widely in primary productivity; add guano or fertilizer to one plot of each pair	Seabirds and marine mammals; terrestrial plants and animals	Primary productivity, rates of decomposition, mineralization, numbers of herbivores
Trees as condensation nuclei	Trees augment water inputs to dry islands exposed to trade winds	Plant trees (e.g., *Pinus canariensis*) on one member of paired watershed or islands that receive similar rainfall; measure rain in clearing and in plantation; measure throughfall and stemflow	Local flora	Hydrology (driven by total water inputs as sum of rain, throughfall, and stemflow)

Table 18.1. (Contd.)

General topic	Prediction	Experiment	Biodiversity components	Ecosystem processes
Mangrove recovery	Species-rich mangrove forests are more resilient than species-poor stands	Kill groups of trees or stands along mangrove diversity gradients (which may be related to typhoon-frequency gradient across islands)	Mangroves and animals dependent on mangrove communities	Colonization and growth rates of mangrove individuals; production of birds, insects, fish, and root epifauna
Mangrove mortality	Individual tree mortality prevails in species-rich mangrove areas, while stand-level dieback or mass mortality prevail in species-poor mangrove areas	Establish large, permanent inventory plots in many mangrove stands that differ with respect to diversity. Await natural mortality, including that caused by episodic agents such as typhoons	Mangrove species diversity; animals dependent on mangrove communities	Mortality and growth of mangrove individuals
Soil erosion	A diversity of plant species increases canopy density, thereby reducing the erosive impact of rain	Create diverse and simple plantations on deforested lands similar in aspect, soil, topography, and rainfall	Plant species, both native and alien	Runoff amount, rate, sediment load; character of erosion (sheet, rill, gully); infiltration, especially as related to litter quality and quantity
Ecological economics of islands	A subsistence index, defined as the ratio between outside "industrial" inputs and the island's internal capacity for self-maintenance, is inversely proportional to the integrity of the native biota	Choose paired islands in the same archipelago that differ, due to historical reasons, in the degree to which the economy is dependent upon outside inputs. Construct energy (and money) budgets of each, and attempt to reconstruct time course of subsidy in relation to status of natural ecosystems on each	Fraction of land covered by vegetation; fraction of land covered by native plants; sea bird colonies	Primary productivity of terrestrial systems; nutrient return from sea to land via biological pumps; detritus flux from land to sea; productivity of inshore marine systems

taxonomic affinity, ecological similarity, and human acceptability. In addition to the impacts of these introductions on ecosystem processes, it would also be informative to measure the functional properties of the substitute, such as its physiology, behavioral ecology, and phenology.

There could be a number of consequences of adding or deleting functional groups. These might include the acceleration, deceleration, or stagnation of succession; increased or decreased resistance, and responsiveness to pest attacks; absolute and relative changes in primary and secondary productivity; and changed vulnerability to fire. Using the degree of habitat specialization among plants, it may be possible to assess the effects of diversity on ecosystem processes through species removal or revegetation experiments.

Many islands are volcanic, and volcanism (especially in island arc systems derived from hot spots) creates sharp gradients in substrate, upon which other environmental factors are relatively uniform. This leads to a unique opportunity for replicated experimentation involving limiting factors and their changes over time. Furthermore, such comparisons can easily be expanded to accommodate comparisons across archipelagos as well as between islands and continents.

18.3.2 Larger-Scale Phenomena

Other island experiments would lead primarily to insights about the relationship between biological diversity and ecosystem processes at the level of the island and the larger land- and sea-scape of which it is a part. For instance, many islands are focal points for sea-to-land transfer of nutrients via birds, which harvest secondary productivity from vast areas of ocean, and deposit guano on land. A first logical step might involve elimination of an exotic predator, such as the feral cats that are so common on islands that once supported huge colonies of seabirds. Such a program should only be undertaken after first considering possible cascading or linked effects: what will become of exotic rat populations in the absence of cats, and what implications does this hold for the island's biota?

Experiments with coastal fringe ecosystems, such as mangroves, might not require inflicting more damage on these already-battered communities. Instead, they could be implemented by taking advantage of manipulations that have already been induced by development, such as urbanization, agriculture, aquaculture, and drainage for mosquito control, or by using manipulations anticipated or proposed for other purposes, such as mangrove silviculture. As both bridges and barriers between land and sea, coastal fringe ecosystems are linked to processes that merit special attention, such as reef eutrophication, storm penetration, and impacts on detrital food chains of estuaries.

Soil erosion is a response of interest in many kinds of biodiversity experiments, but it is of special significance on islands because of their high edge-to-surface ratio; its potential impact on the inshore marine systems on which the island's human inhabitants may depend for food; and the fact that islands seem to suffer disproportionately from devegetation and the soil erosion it provokes. Biologists should not overlook the economically useful and visible role that native species' conservation and reintroduction can play in controlling erosion.

It may also be useful to scale up from biological systems to the level of ecological economics, for nowhere is the interdependency between humans and their surroundings any tighter than on islands (Defoe 1862). To accommodate this perspective the concept of biodiversity might have to be expanded to include the diversity of human enterprises. Likewise, the concept of an "experiment" might have to be relaxed, although this does not make the research any less crucial, just more difficult. Ecosystem processes, in this case, might imply inputs and outflows tied to human activity, such as economic subsidies, transportation and marketing, capture and use of freshwater, waste disposal, and harvesting (both commercial and subsistence) of products from the surrounding marine ecosystem.

18.4 Conclusions

Well-known, relatively species-poor biota, coupled with well-defined environments not confounded by drastic cross-gradient changes in other factors, make islands excellent settings for experiments on the relationship between biological diversity and ecosystem processes. The main attractions of islands for ecological research are likely to remain the unplanned experiments and island-continent comparisons laid out by their singular environment and unique biogeographical history.

Nevertheless, the superimposition of investigator-imposed treatments on an island's matrix of natural experiments would be a powerful lever in maximizing information yield. Investigators could regard subdivisions of island gradients as experimental blocks, to which further experimental treatments can profitably be added. Islands, as mesocosms of larger-scale phenomena, provide unique, straightforward combinations of organisms and environment, and much can be learned by their manipulation.

Reference

Defoe D (1862) The adventures of Robinson Crusoe. S O Beeton, London

Subject Index

Ecological Studies

Volumes published since 1989

Volume 77
Air Pollution and Forest Decline: A Study of Spruce (*Picea abies*) on Acid Soils (1989)
E.-D. Schulze, O. L. Lange, and R. Oren (Eds.)

Volume 78
Agroecology: Researching the Ecological Basis for Sustainable Agriculture (1990)
S. R. Gliessman (Ed.)

Volume 79
Remote Sensing of Biosphere Functioning (1990)
R. J. Hobbs and H. A. Mooney (Eds.)

Volume 80
Plant Biology of the Basin and Range (1990)
B. Osmond, G. M. Hidy, and L. Pitelka (Eds.)

Volume 81
Nitrogen in Terrestrial Ecosystems: Questions of Productivity, Vegetational Changes, and Ecosystem Stability (1991)
C. O. Tamm

Volume 82
Quantitative Methods in Landscape Ecology: The Analysis and Interpretation of Landscape Heterogeneity (1990)
M. G. Turner and R. H. Gardner (Eds.)

Volume 83
The Rivers of Florida (1990)
R. J. Livingston (Ed.)

Volume 84
Fire in the Tropical Biota: Ecosystem Processes and Global Challenges (1990)
J. G. Goldammer (Ed.)

Volume 85
The Mosaic-Cycle Concept of Ecosystems (1991)
H. Remmert (Ed.)

Volume 86
Ecological Heterogeneity (1991)
J. Kolasa and S. T. A. Pickett (Eds.)

Volume 87
Horses and Grasses: The Nutritional Ecology of Equids and Their Impact on the Camargue (1992)
P. Duncan

Volume 88
Pinnipeds and El Niño: Responses to Environmental Stress (1992)
F. Trillmich and K. A. Ono (Eds.)

Volume 89
Plantago: A Multidisciplinary Study (1992)
P. J. C. Kuiper and M. Bos (Eds.)

Volume 90
Biogeochemistry of a Subalpine Ecosystem: Loch Vale Watershed (1992)
J. Baron (Ed.)

Volume 91
Atmospheric Deposition and Forest Nutrient Cycling (1992)
D. W. Johnson and S. E. Lindberg (Eds.)

Volume 92
Landscape Boundaries: Consequences for Biotic Diversity and Ecological Flows (1992)
A. J. Hansen and F. di Castri (Eds.)

Volume 93
Fire in South African Mountain Fynbos: Ecosystem, Community, and Species Response at Swartboskloof (1992)
B. W. van Wilgen et al. (Eds.)

Volume 94
The Ecology of Aquatic Hyphomycetes (1992)
F. Bärlocher (Ed.)

Volume 95
Palms in Forest Ecosystems of Amazonia (1992)
F. Kahn and J.-J. DeGranville

Volume 96
Ecology and Decline of Red Spruce in the Eastern United States (1992)
C. Eagar and M. B. Adams (Eds.)

Ecological Studies

Volumes published since 1989

Volume 97
The Response of Western Forests to Air Pollution (1992)
R.K. Olson, D. Binkley, and M. Böhm (Eds.)

Volume 98
Plankton Regulation Dynamics (1993)
N. Walz (Ed.)

Volume 99
Biodiversity and Ecosystem Function (1993)
E.-D. Schulze and H.A. Mooney (Eds.)

Volume 100
Ecophysiology of Photosynthesis (1994)
E.-D. Schulze and M.M. Caldwell (Eds.)

Volume 101
Effects of Land Use Change on Atmospheric CO_2 Concentrations: South and South East Asia as a Case Study (1993)
V.H. Dale (Ed.)

Volume 102
Coral Reef Ecology (1993)
Y.I. Sorokin

Volume 103
Rocky Shores: Exploitation in Chile and South Africa (1993)
W.R. Siegfried (Ed.)

Volume 104
Long-Term Experiments With Acid Rain in Norwegian Forest Ecosystems (1993)
G. Abrahamsen et al. (Eds.)

Volume 105
Microbial Ecology of Lake Plußsee (1993)
J. Overbeck and R.J. Chrost (Eds.)

Volume 106
Minimum Animal Populations (1994)
H. Remmert (Ed.)

Volume 107
The Role of Fire in Mediterranean-Type Ecosystems (1994)
J.M. Moreno and W.C. Oechel

Volume 108
Ecology and Biogeography of Mediterranean Ecosystems in Chile, California and Australia (1994)
M.T.K. Arroyo, P.H. Zedler, and M.D. Fox (Eds.)

Volume 109
Mediterranean-Type Ecosystems. The Function of Biodiversity (1995)
G.W. Davis and D.M. Richardson (Eds.)

Volume 110
Tropical Montane Cloud Forests (1995)
L.S. Hamilton, J.O. Juvik, and F.N. Scatena (Eds.)

Volume 111
Peatland Forestry. Ecology and Principles (1995)
E. Paavilainen and J. Päivänen

Volume 112
Tropical Forests: Management and Ecology (1995)
A.E. Lugo and C. Lowe (Eds.)

Volume 113
Arctic and Alpine Biodiversity. Patterns, Causes and Ecosystem Consequences (1995)
F.S. Chapin III and C. Körner (Eds.)

Volume 114
Crassulacean Acid Metabolism. Biochemistry, Ecophysiology and Evolution (1995)
K. Winter and J.A.C. Smith (Eds.)

Volume 115
Islands. Biological Diversity and Ecosystem Function (1995)
P.M. Vitousek, L.L. Loope, and H. Adsersen (Eds.)

Volume 116
High Latitude Rainforests and Associated Ecosystems of the West Coast of the Americas: Climate, Hydrology, Ecology and Conservation (1995)
R.G. Lawford, P. Alaback, and E.R. Fuentes (Eds.)

Volume 117
Global Change and Mediterranean-Type Ecosystems (1995)
J. Moreno and W.C. Oechel (Eds.)

Springer-Verlag and the Environment

MIX
Papier aus verantwortungsvollen Quellen
Paper from responsible sources
FSC® C105338

If you have any concerns about our products,
you can contact us on
ProductSafety@springernature.com

In case Publisher is established outside the EU,
the EU authorized representative is:
Springer Nature Customer Service Center GmbH
Europaplatz 3, 69115 Heidelberg, Germany

Printed by Libri Plureos GmbH
in Hamburg, Germany